Ben Stacy Jerrik (Ed.)

Forest Hills, Pennsylvania

Ben Stacy Jerrik (Ed.)

Forest Hills, Pennsylvania

Allegheny County, Pennsylvania, 2010 United States Census, United States Census Bureau, Braddock Hills, Pennsylvania

Part Press

Imprint

All parts of this book are extracted from Wikipedia, the free encyclopedia (www.wikipedia.org).

You can get detailed informations about the authors of this collection of articles at the end of this book. The editors (Ed.) of this book are no authors. They have not modified or extended the original texts.

Pictures published in this book can be under different licences than the GNU Free Documentation License. You can get detailed informations about the authors and licences of pictures at the end of this book.

The content of this book was generated collaboratively by volunteers. Please be advised that nothing found here has necessarily been reviewed by people with the expertise required to provide you with complete, accurate or reliable information. Some information in this book maybe misleading or wrong. The Publisher does not guarantee the validity of the information found here. If you need specific advice (f.e. in fields of medical, legal, financial, or risk management questions) please contact a professional who is licensed or knowledgeable in that area.

Cover image: www.ingimage.com
Concerning the licence of the cover image please contact ingimage.

Publisher:
Part Press is a trademark of
International Book Market Service Ltd., 17 Rue Meldrum, Beau Bassin, 1713-01 Mauritius
Email: info@bookmarketservice.com
Website: www.bookmarketservice.com

Published in 2013

Printed in: U.S.A., U.K., Germany. This book was not produced in Mauritius.

ISBN: 978-620-1-57980-4

Contents

Articles

References

Trois_novelettes_(Poulenc)

Trois novelettes (*Three novelettes*) are three short pieces for piano composed by Francis Poulenc.

The first two novelettes, in C major and B-flat minor, written in 1927 and 1928 respectively, were originally published together, with the third novelette in E minor written in 1959. These novelettes demonstrate multi-layered piano writing.

- Novelette in C major (1927) is dedicated to Virginie Liénard, a family friend, also the dedicatee of his piano suite *Les soirées de Nazelles* (1930-36). The piece features a neo-classical song-like melody interrupted by a central contrasting section.
- Novelette in B-flat minor (1928) is similar in ways to a scherzo, dedicated to Poulenc's friend and critic, Louis Laloy.
- Novelette in E minor (1959) is based on Manuel de Falla's 7/8 theme from *El amor brujo* (1916), simplified by Poulenc in 3/8 time, without a contrasting section. This piece was dedicated to his close friend R. Douglas Gibson, employing fluid arpeggio figures.

Poulenc premiered his first two novelettes at a concert in Paris on June 10, 1928.

References

Sachania, M. & Potter, C. *Francis Poulenc: Three Novelettes for Piano*, Chester Music, revised edition (1999).

Novelette

Novelette may refer to:

- a novella, especially with trivial or sentimental themes
- a narrative prose fiction shorter than a novella and longer than a short story
  - Hugo Award for Best Novelette
  - Nebula Award for Best Novelette
- novelette (music), short piece of lyrical music
- Novelette, protagonist of *Da Kink in My Hair*, 2001 play by Trey Anthony

__DISAMBIG__

Piano

The **piano** is a musical instrument played by means of a keyboard. Widely used in classical and jazz music for solo performances, ensemble use, chamber music and accompaniment, the piano is also popular as a tool for composing and rehearsal. Although not portable and often expensive, the piano's versatility and ubiquity have made it one of the world's most familiar musical instruments.

Pressing a key on the piano's keyboard causes a padded (often with felt) hammer to strike steel strings. The hammers rebound, and the strings continue to vibrate at their resonant frequency.[1] These vibrations are transmitted through a bridge to a sounding board that more efficiently couples the acoustic energy to the air. The sound would otherwise be no louder than that directly produced by the strings. When the key is released, a damper stops the string's vibration and the sound. See the article on Piano key frequencies for a picture of the piano keyboard and the location of middle-C. In the Hornbostel-Sachs system of instrument classification, pianos are considered chordophones.

The word *piano* is a shortened form of ***pianoforte*** (PF), the Italian word for the instrument (which in turn derives from the previous terms *gravicembalo col piano e forte* and *fortepiano*). The Italian musical terms *piano* and *forte* indicate "soft" and "strong" respectively,[2] in this context referring to the variations in sound volume the instrument produces in response to a pianist's touch on the keys: the greater the velocity of a key press, the greater the force of the hammer hitting the strings, and the louder the sound of the note produced.

History

The piano was founded on earlier technological innovations. The first string instruments with struck strings were the hammered dulcimers.[3] During the Middle Ages, there were several attempts at creating stringed keyboard instruments with struck strings.[4] By the 17th century, the mechanisms of keyboard instruments such as the clavichord and the harpsichord were well known. In a clavichord the strings are struck by tangents, while in a harpsichord they are plucked by quills. Centuries of work on the mechanism of the harpsichord in particular had shown the most effective ways to construct the case, soundboard, bridge, and keyboard for a mechanism intended to hammer strings.

Grand piano by Louis Bas of Villeneuve-lès-Avignon, France, 1781. Earliest French grand piano known to survive; includes an inverted wrestplank and action derived from the work of Bartolomeo Cristofori (ca. 1700) with ornately decorated soundboard.

The invention of the modern piano is credited to Bartolomeo Cristofori (1655–1731) of Padua, Italy, who was employed by Ferdinando de' Medici, Grand Prince of Tuscany, as the Keeper of the Instruments. He was an expert harpsichord maker, and was well acquainted with the body of knowledge on stringed keyboard instruments. It is not known exactly when Cristofori first built a piano. An inventory made by his employers, the Medici family, indicates the existence of a piano by the year 1700; another document of doubtful authenticity indicates a date of 1698. The three Cristofori pianos that survive today date from the 1720s.[5] [6]

While the clavichord allowed expressive control of volume and sustain, it was too quiet for large performances. The harpsichord produced a sufficiently loud sound, but had little expressive control over each note. The piano was likely formed as an attempt to combine loudness with control, avoiding the trade-offs of available instruments.

Cristofori's great success was solving, with no prior example, the fundamental mechanical problem of piano design: the hammer must strike the string, but not remain in contact with it (as a tangent remains in contact with a clavichord string) because this would dampen the sound. Moreover, the hammer must return to its rest position without bouncing violently, and it must be possible to repeat a note rapidly. Cristofori's piano action was a model for the many different approaches to piano actions that followed. Cristofori's early instruments were made with thin strings, and were much quieter than the modern piano—but compared to the clavichord (the only previous keyboard instrument capable of dynamic nuance via the keyboard) they were much louder and had more sustain.

Early piano replica by the modern builder Paul McNulty, after Walter & Sohn, 1805

Cristofori's new instrument remained relatively unknown until an Italian writer, Scipione Maffei, wrote an enthusiastic article about it in 1711, including a diagram of the mechanism. This article was widely distributed, and most of the next generation of piano builders started their work due to reading it. One of these builders was Gottfried Silbermann, better known as an organ builder. Silbermann's pianos were virtually direct copies of Cristofori's, with one important addition: Silbermann invented the forerunner of the modern damper pedal, which lifts all the dampers from the strings simultaneously.

Silbermann showed Johann Sebastian Bach one of his early instruments in the 1730s, but Bach did not like it at that time, claiming that the higher notes were too soft to allow a full dynamic range. Although this earned him some animosity from Silbermann, the criticism was apparently heeded. Bach did approve of a later instrument he saw in 1747, and even served as an agent in selling Silbermann's pianos.[7]

Piano-making flourished during the late 18th century in the Viennese school, which included Johann Andreas Stein (who worked in Augsburg, Germany) and the Viennese makers Nannette Streicher (daughter of Stein) and Anton Walter. Viennese-style pianos were built with wood frames, two strings per note, and had leather-covered hammers. Some of these Viennese pianos had the opposite coloring of modern-day pianos; the natural keys were black and the accidental keys white.[] It was for such instruments that Wolfgang Amadeus Mozart composed his concertos and sonatas, and replicas of them are built today for use in authentic-instrument performance of his music. The pianos of Mozart's day had a softer, more ethereal tone than today's pianos or English pianos, with less sustaining power. The term *fortepiano* is now used to distinguish the 18th-century instrument from later pianos.

In the period lasting from about 1790 to 1860, the Mozart-era piano underwent tremendous changes that led to the modern form of the instrument. This revolution was in response to a preference by composers and pianists for a more powerful, sustained piano sound, and made possible by the ongoing Industrial Revolution with resources such as high-quality piano wire for strings, and precision casting for the production of iron frames. Over time, the tonal range of the piano was also increased from the five octaves of Mozart's day to the 7⅓ or more octaves found on modern pianos.

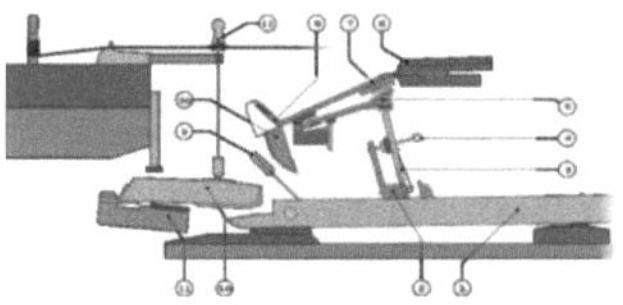

Broadwood square action (click for page with legend)

Early technological progress owed much to the firm of Broadwood. John Broadwood joined with another Scot, Robert Stodart, and a Dutchman, Americus Backers, to design a piano in the harpsichord case—the origin of the "grand". They achieved this in about 1777. They quickly gained a reputation for the splendour and powerful tone of their instruments, with Broadwood constructing ones that were progressively larger, louder, and more robustly constructed. They sent pianos to both Joseph Haydn and Ludwig van Beethoven, and were the first firm to build pianos with a range of more than five octaves: five octaves and a fifth during the 1790s, six octaves by 1810 (Beethoven used the extra notes in his later works), and seven octaves by 1820. The Viennese makers similarly followed these trends; however the two schools used different piano actions: Broadwoods were more robust, Viennese instruments were more sensitive.

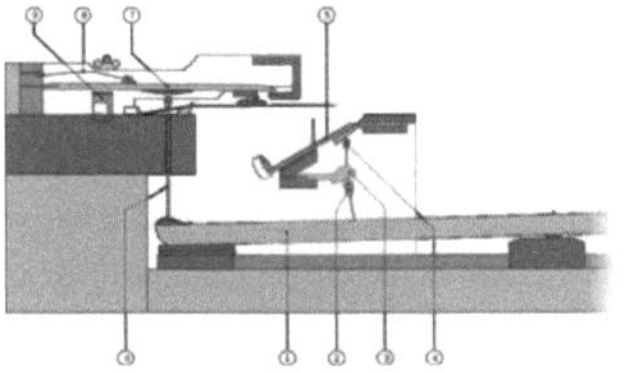

Erard square action (click for page with legend)

By the 1820s, the center of innovation had shifted to Paris, where the Pleyel firm manufactured pianos used by Frédéric Chopin and the Érard firm manufactured those used by Franz Liszt. In 1821, Sébastien Érard invented the double escapement action, which incorporated a *repetition lever* (also called the *balancier*) that permitted repeating a note even if the key had not yet risen to its maximum vertical position. This facilitated rapid playing of repeated notes, a musical device pioneered by Liszt. When the invention became public, as revised by Henri Herz, the double escapement action gradually became standard in grand pianos, and is still incorporated into all grand pianos currently produced.

Other improvements of the mechanism included the use of felt hammer coverings instead of layered leather or cotton. Felt, which was first introduced by Henri Pape in 1826, was a more consistent material, permitting wider dynamic ranges as hammer weights and string tension increased. The sostenuto pedal (see below), invented in 1844 by Jean-Louis Boisselot and copied by the Steinway firm in 1874, allowed a wider range of effects.

One innovation that helped create the sound of the modern piano was the use of a strong iron frame. Also called the "plate", the iron frame sits atop the soundboard, and serves as the primary bulwark against the force of string tension. The increased structural integrity of the iron frame allowed the use of thicker, tenser, and more numerous strings. In a modern grand the total string tension can exceed 20 tons. The single piece cast iron frame was patented in 1825 in Boston by Alpheus Babcock, combining the metal hitch pin plate (1821, claimed by Broadwood on behalf of Samuel Hervé) and resisting bars (Thom and Allen, 1820, but also claimed by Broadwood and Érard). Babcock later worked for the Chickering & Mackays firm who patented the first full iron frame for grand pianos in 1843. Composite forged metal frames were preferred by many European makers until the American system was fully adopted by the early 20th century.

Other important advances included changes to the way the piano is strung, such as the use of a "choir" of three strings rather than two for all but the lowest notes, and the implementation of an over-strung scale, in which the strings are placed in two separate planes, each with its own bridge height. (This is also called *cross-stringing*. Whereas earlier instruments' bass strings were a mere continuation of a single string plane, over-stringing placed the bass bridge behind and to the treble side of the tenor bridge area. This *crossed* the strings, with the bass strings in the higher plane.) This permitted a much narrower cabinet at the "nose" end of the piano, and optimized the transition from unwound tenor strings to the iron or copper-wrapped bass strings. Over-stringing was invented by Jean-Henri Pape during the 1820s, and first patented for use in grand pianos in the United States by Henry Steinway, Jr. in 1859.

Duplex scaling of an 1883 Steinway Model 'A'. From lower left to upper right: main sounding length of strings, treble bridge, duplex string length, duplex bar (nickel-plated bar parallel to bridge), hitchpins, plate strut with bearing bolt, plate hole.

Some piano makers developed schemes to enhance the tone of each note. Julius Blüthner developed Aliquot stringing in 1893 as well as Pascal-Joseph Taskin [8] (1788), and Collard & Collard (1821). Each used more distinctly ringing, undamped vibrations to modify tone, except the Blüthner Aliquot stringing, which uses an additional fourth string in the upper two treble sections. While the hitchpins of these separately suspended Aliquot strings are raised slightly above the level of the usual tri-choir strings, they are not struck by the hammers but rather are damped by attachments of the usual dampers. Eager to copy these effects, Theodore Steinway invented *duplex scaling*, which used short lengths of non-speaking wire bridged by the aliquot throughout much of upper range of the piano, always in locations that caused them to vibrate in conformity with their respective overtones—typically in doubled octaves and twelfths

Some early pianos had shapes and designs that are no longer in use. The square piano (not truly square, but rectangular) was cross strung at an extremely acute angle above the hammers, with the keyboard set along the long side. This design is attributed to Gottfried Silbermann or Christian Ernst Friderici on the continent, and Johannes Zumpe or Harman Vietor in England, and it was improved by changes first introduced by Guillaume-Lebrecht Petzold in France and Alpheus Babcock in the United States. Square pianos were built in great numbers through the 1840s in Europe and the 1890s in America, and saw the most visible change of any type of piano: the iron-framed, over-strung squares manufactured by Steinway & Sons were more than two-and-a-half times the size of Zumpe's wood-framed instruments from a century before. Their overwhelming popularity was due to inexpensive construction and price, although their tone and performance were limited by narrow soundboards, simple actions and string spacing that made proper hammer alignment difficult.

The mechanism and strings in upright pianos are perpendicular to the keys.

The tall, vertically strung upright grand was arranged like a grand set on end, with the soundboard and bridges above the keys, and tuning pins below them. The term was later revived by many manufacturers for advertising purposes. Giraffe, pyramid and lyre pianos were arranged in a somewhat similar fashion in evocatively shaped cases.

The very tall cabinet piano was introduced about 1805 and was built through the 1840s. It had strings arranged vertically on a continuous frame with bridges extended nearly to the floor, behind the keyboard and very large *sticker action*. The short cottage upright or *pianino* with vertical stringing, made popular by Robert Wornum around 1815, was built into the 20th century. They are informally called *birdcage pianos* because of their prominent damper mechanism. Pianinos were distinguished from the oblique, or diagonally strung upright made popular in France by Roller & Blanchet during the late 1820s. The tiny spinet upright was manufactured from the mid-1930s until recent times. The low position of the hammers required the use of a "drop action" to preserve a reasonable keyboard height.

Modern upright and grand pianos attained their present forms by the end of the 19th century. Improvements have been made in manufacturing processes, and many individual details of the instrument continue to receive attention.

History and musical performance

Thelonious Monk at Minton's Playhouse, New York City, 1947

Much of the most widely admired piano repertoire in classical music, for example, that of Haydn, Mozart, and Beethoven, was composed for a type of instrument (the fortepiano) that is rather different from modern instruments this music is normally performed on today. Even the music of the Romantic movement, including Liszt, Chopin, Robert Schumann, Felix Mendelssohn and Johannes Brahms, was written for pianos substantially different from modern pianos.

Starting in Beethoven's later career, the fortepiano evolved into the modern piano as we know it today. Modern pianos were in wide use by the late 19th century. They featured an octave range larger than the earlier fortepiano instrument, adding around 30 more keys to the instrument. The mechanical action structure of the upright piano was invented in London, England in 1826 by Robert Wornum, and upright models became the most popular model, also amplifying the sound.[9] Mass production in factories made them more affordable for a larger number of people. They appeared in music halls and pubs during the 19th century, providing entertainment through a piano soloist, or in combination with a small band. Pianists began accompanying singers or dancers performing on stage, or patrons dancing on a dance floor.

During the 19th century, American musicians playing for working-class audiences in small pubs and bars, particularly African-American composers, developed new musical genres based on the modern piano. Ragtime music, popularized by composers such as Scott Joplin, reached a broader audience by 1900. The popularity of ragtime music was quickly succeeded by Jazz piano. New techniques and rhythms were invented for the piano, including ostinato for boogie woogie, and Shearing voicing. George Gershwin's Rhapsody in Blue broke new musical ground by combining American jazz piano with symphonic sounds. Comping, a technique for accompanying jazz vocalists on piano, was exemplified by Duke Ellington's technique. Honky tonk music, featuring yet another style of piano rhythm, became popular during the same era. Bebop techniques grew out of jazz, with leading composers such as Thelonious Monk and Bud Powell. In the late 20th century, Bill Evans composed pieces combining classical techniques with his jazz experimentations. Herbie Hancock was one of the first jazz pianists to find mainstream popularity working with newer urban music techniques.

Pianos have also been used in rock and roll by entertainers such as Jerry Lee Lewis, Little Richard, Elton John, and Billy Joel to name just a few.

Modernist styles of music have also appealed to composers writing for the modern grand piano, including John Cage and Philip Glass.

Modern piano

A schematic depiction of the construction of a pianoforte

frame (**1**)
lid, front part (**2**)
capo bar (**3**)
damper (**4**)
lid, back part (**5**)
damper mechanism (**6**)
sostenuto rail (**7**)
pedal mechanism, rods (**8, 9, 10**)
pedals: right (sustain/damper), middle (sostenuto), left (soft/una-corda) (**11**)
bridge (**12**)
hitch pin (**13**)
frame (**14**)
sound board (**15**)
string (**16**)

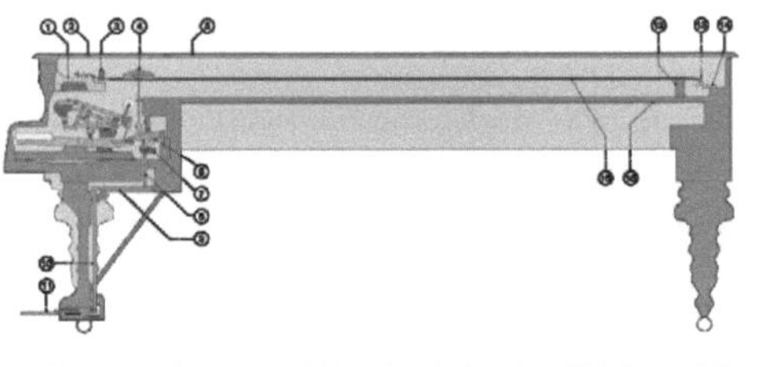

Physics

When the key is struck, a chain reaction occurs to produce the sound. First, the key raises the wippen, which forces the jack against the hammer roller (or "knuckle"). The hammer roller then lifts the lever carrying the hammer. The key also raises the damper; and immediately after the hammer strikes the wire it falls back, allowing the wire to resonate. When the key is released the damper falls back onto the strings, stopping the wire from vibrating.[10] The vibrating piano strings themselves are not very loud, but their vibrations are transmitted to a large soundboard that moves air and thus converts the energy to sound. The irregular shape and off-center placement of the bridge ensures that the soundboard vibrates strongly at all frequencies.[11] (See Piano action for a diagram and detailed description of piano parts.)

There are three factors that influence the pitch of a vibrating wire.

- Length: All other factors the same, the shorter the wire, the higher the pitch.
- Mass per unit length: All other factors the same, the thinner the wire, the higher the pitch.
- Tension: All other factors the same, the tighter the wire, the higher the pitch.

A vibrating wire subdivides itself into many parts vibrating at the same time. Each part produces a pitch of its own, called a partial. A vibrating string has one fundamental and a series of partials. The most pure combination of two pitches is when one is double the frequency of the other.[12]

For a repeating wave, the velocity, v, equals the wavelength, λ, times the frequency, f.

$$v = \lambda f$$

On the piano string, waves reflect from both ends. The superposition of reflecting waves results in a standing wave pattern, but only for wavelengths $\lambda = 2L, L, L/2, \ldots = 2L/n$, where L is the length of the string. Therefore the only frequencies produced on a single string are $f = nv/(2L)$. Timbre is largely determined by the content of these harmonics. Different instruments have different harmonic content for the same pitch. A real string vibrates at harmonics that are not perfect multiples of the fundamental. This results in a little inharmonicity, which gives richness to the tone but causes significant tuning challenges throughout the compass of the instrument.[11]

Striking the piano key with greater velocity increases the amplitude of the waves and therefore the volume. From *pianissimo* (*pp*) to *fortissimo* (*ff*) the hammer velocity changes by almost a factor of a hundred. The hammer contact time with the string shortens from 4 ms at *pp* to less than 2 ms at *ff*.[11] If two wires adjusted to the same pitch are struck at the same time, the sound produced by one reinforces the other, and a louder combined sound of shorter duration is produced. If one wire vibrates out of synchronization with the other, they subtract from each other and produce a softer tone of longer duration.[13]

Types

Modern pianos have two basic configurations (with subcategories): the grand piano and the upright piano.

Grand

Steinway grand piano in the White House

August Förster upright piano

In grand pianos, the frame and strings are horizontal, with the strings extending away from the keyboard. The action lies beneath the strings, and uses gravity as its means of return to a state of rest.

There are many sizes of grand piano. A rough generalization distinguishes the *concert grand* (between 2.2 and 3 metres long, about –feet) from the *parlor grand* or *boudoir grand* (1.7 to 2.2 metres long, about –feet) and the smaller *baby grand* (around 1.5 metres (feet)).

All else being equal, longer pianos with longer strings have larger, richer sound and lower inharmonicity of the strings. Inharmonicity is the degree to which the frequencies of overtones (known as partials or harmonics) sound sharp relative to whole multiples of the fundamental frequency. This results from the piano's considerable string stiffness; as a struck string decays its harmonics vibrate, not from their termination, but from a point very slightly toward the center (or more flexible part) of the string. The higher the partial, the further sharp it runs. Pianos with shorter and thicker string (i.e. small pianos with short string scales) have more inharmonicity. The greater the inharmonicity, the more the ear perceives it as harshness of tone.

Inharmonicity requires that octaves be *stretched,* or tuned to a lower octave's corresponding sharp overtone rather than to a theoretically correct octave. If octaves are not stretched, single octaves sound in tune, but double—and notably triple—octaves are unacceptably narrow. Stretching a small piano's octaves to match its inherent inharmonicity level creates an imbalance among all the instrument's intervallic relationships, not just its octaves. In a concert grand, however, the octave "stretch" retains harmonic balance, even when aligning treble notes to a harmonic produced from three octaves below. This lets close and widespread octaves sound pure, and produces virtually beatless perfect fifths. This gives the concert grand a brilliant, singing and sustaining tone quality—one of the principal reasons that full-size grands are used in the concert hall. Smaller grands satisfy the space and cost needs of domestic use.

Upright (vertical)

Upright pianos, also called vertical pianos, are more compact because the frame and strings are vertical. The hammers move horizontally, and return to their resting position via springs, which are susceptible to degradation. Upright pianos with unusually tall frames and long strings are sometimes called *upright grand* pianos. Some authors classify modern pianos according to their height and to modifications of the action that are necessary to accommodate the height.

- *Studio* pianos are around 42 to 45 inches tall. This is the shortest cabinet that can accommodate a full-sized action located above the keyboard.
- *Console* pianos have a compact action (shorter hammers), and are a few inches shorter than studio models.
- The top of a *spinet* model barely rises above the keyboard. The action is located below, operated by vertical wires that are attached to the backs of the keys.
- Anything taller than a studio piano is called an *upright*.

Other types

Player piano from 1920 (Steinway)

The toy piano was introduced in the 19th century.

In 1863, Henri Fourneaux invented the player piano, which plays itself from a piano roll. A machine perforates a performance recording into rolls of paper, and the player piano replays the performance using pneumatic devices. Modern equivalents of the player piano include the Bösendorfer CEUS and the Yamaha disklavier, using solenoids and MIDI rather than pneumatics and rolls.

A silent piano is an acoustic piano having an option to silence the strings by means of an interposing hammer bar. They are designed for private silent practice.

Edward Ryley invented the transposing piano in 1801. It has a lever under the keyboard as to move the keyboard relative to the strings so a pianist can play in a familiar key while the music sounds in a different key.

The prepared piano, present in some contemporary art music, is a piano with objects placed inside it to alter its sound, or has had its mechanism changed in some other way. The scores for music for prepared piano specify the modifications, for example instructing the pianist to insert pieces of rubber, paper, metal screws, or washers in between the strings. These either mute the strings or alter their timbre. A harpsichord-like sound can be produced by placing or dangling small metal buttons in front of the hammer.

In 1954 a German company exhibited a wire-less piano at the Spring Fair in Frankfurt, Germany that sold for $238. The wires were replaced by metal bars of different alloys that replicated the standard wires when played.[14] A similar concept is used in the electric-acoustic Rhodes piano.

Electric pianos use electromagnetic pickups to amplify the sound of the strings. Playing a note loudly causes the electric signal to clip, and the player can incorporate the distortion into his or her expressive range.

Digital pianos use digital sampling technology to reproduce the sound of each piano note. Digital pianos can be sophisticated, with features including working pedals, weighted keys, multiple voices, and MIDI interfaces. However, when one depresses the damper pedal (see below) on such an instrument, there are no strings to vibrate sympathetically. The synthesis software of some higher end digital pianos, such as the Yamaha Clavinova series, or the Kawai MP8 series, incorporates physical models of sympathetic vibration.

With the advent of powerful desktop computers, highly realistic pianos have become available as affordable software modules. Some of these modules use multi-gigabyte piano sample sets with as many as 90 recordings, each lasting many seconds, for each of the 88 (some have 81) keys under different conditions. Additional samples emulate sympathetic resonance, key release, the drop of the dampers, and simulations of piano techniques like re-pedaling to augment these conditions. Some other software modules, such as Modartt's 2006 Pianoteq, use no samples whatsoever and are a pure synthesis of all aspects of the physicalities that go into the creation of a real piano's sound.

Today, piano manufactures take advantage of innovative pianos that play themselves via a CD or MP3 player. Similar in concept to a player piano, the PianoDisc or iQ systems allow pianos to "play themselves" when the software interprets a certain file format. Such additions are quite expensive, often doubling the cost of a piano. These pianos are available in both upright and grand.

Keyboard

Almost every modern piano has 52 white keys and 36 black keys for a total of 88 keys (seven octaves plus a minor third, from A_0 to C_8). Many older pianos only have 85 keys (seven octaves from A_0 to A_7). Some piano manufacturers extend the range further in one or both directions.

Keyboard of a grand piano

Some Bösendorfer pianos, for example, extend the normal range down to F_0, and one of their models even goes as far as a bottom C_0, making a full eight octave range. These extra keys are sometimes hidden under a small hinged lid that can cover the keys to prevent visual disorientation for pianists unfamiliar with the extra keys, or the colors of the extra white keys are reversed (black instead of white).

The extra keys are added primarily for increased resonance from the associated strings; that is, they vibrate sympathetically with other strings whenever the damper pedal is depressed and thus give a fuller tone. Only a very small number of works composed for piano actually use these notes. More recently, the Stuart and Sons company has also manufactured extended-range pianos, with the first 102 key piano. On their instruments, the frequency range extends from C_0 to F_8, which is the widest practical range for the acoustic piano. The extra keys are the same as the other keys in appearance.

Small studio upright acoustical pianos with only 65 keys have been manufactured for use by roving pianists. Known as *gig* pianos and still containing a cast iron harp (frame), these are comparatively lightweight and can be easily transported to and from engagements by only two people. As their harp is longer than that of a spinet or console piano, they have a stronger bass sound that to some pianists is well worth the trade-off in range that a reduced key-set offers.

The toy piano manufacturer Schoenhut started manufacturing both grands and uprights with only 44 or 49 keys, and shorter distance between the keyboard and the pedals. These pianos are true pianos with action and strings. The pianos were introduced to their product line in response to numerous requests in favor of it.

Pianos have been built with alternative keyboard systems, e.g., the Jankó keyboard.

Pedals

Standard pedals

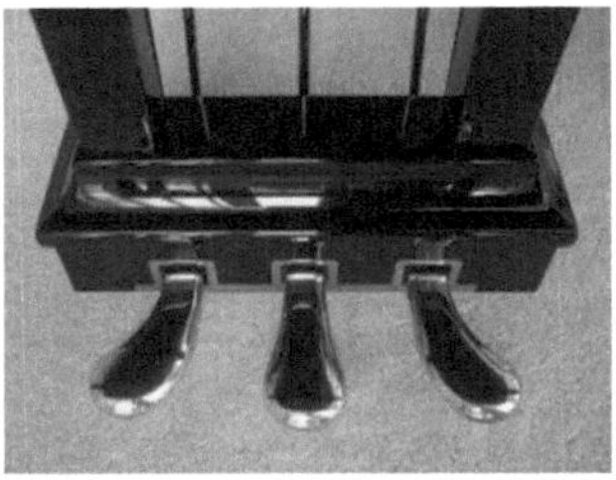

Piano pedals from left to right: una corda, sostenuto and sustain pedal

Pianos have had pedals, or some close equivalent, since the earliest days. (In the 18th century, some pianos used levers pressed upward by the player's knee instead of pedals.) Most grand pianos in the US have three pedals: the soft pedal (una corda), sostenuto, and sustain pedal (from left to right, respectively), while in Europe, the standard is two pedals: the soft pedal and the sustain pedal. Most modern upright pianos also have three pedals: soft pedal, practice pedal and sustain pedal, though older or cheaper models may lack the practice pedal. In Europe the standard for upright pianos is two pedals: the soft and the sustain pedals.

The sustain pedal (or, damper pedal) is often simply called "the pedal", since it is the most frequently used. It is placed as the rightmost pedal in the group. It lifts the dampers from all keys, sustaining all played notes. In addition, it alters the overall tone by allowing all strings, including those not directly played, to reverberate.

The soft pedal or *una corda* pedal is placed leftmost in the row of pedals. In grand pianos it shifts the entire action/keyboard assembly to the right (a very few instruments have shifted left) so that the hammers hit two of the three strings for each note. In the earliest pianos whose unisons were bichords rather than trichords, the action shifted so that hammers hit a single string, hence the name *una corda*, or 'one string'. The effect is to soften the note as well as change the tone. In uprights this action is not possible; instead the pedal moves the hammers closer to the strings, allowing the hammers to strike with less kinetic energy. This produces a slightly softer sound, but no change in timbre.

On grand pianos, the middle pedal is a sostenuto pedal. This pedal keeps raised any damper already raised at the moment the pedal is depressed. This makes it possible to sustain selected notes (by depressing the sostenuto pedal before those notes are released) while the player's hands are free to play additional notes (which aren't sustained). This can be useful for musical passages with pedal points and other otherwise tricky or impossible situations.

On many upright pianos, the middle pedal is called the "practice" or *celeste* pedal. This drops a piece of felt between the hammers and strings, greatly muting the sounds. This pedal can be shifted while depressed, into a "locking" position.

There are also non-standard variants. On some pianos (grands and verticals), the middle pedal can be a bass sustain pedal: that is, when it is depressed, the dampers lift off the strings only in the bass section. Players use this pedal to sustain a single bass note or chord over many measures, while playing the melody in the treble section. On the Stuart and Sons piano as well as the largest Fazioli piano, there is a fourth pedal to the left of the principal three. This fourth pedal works in the same way as the soft pedal of an upright piano, moving the hammers closer to the strings.[15]

Unusual pedals

The rare transposing piano (an example of which was owned by Irving Berlin) has a middle pedal that functions as a clutch that disengages the keyboard from the mechanism, so the player can move the keyboard to the left or right with a lever. This shifts the entire piano action so the pianist can play music written in one key so that it sounds in a different key.

An upright pedal piano by Challen

Some piano companies have included extra pedals other than the standard two or three. Crown and Schubert Piano Co. produced a four-pedal piano. Fazioli currently offers a fourth pedal that provides a second soft pedal, that works by bringing the keys closer to the strings.

Wing and Son of New York offered a five-pedal piano from approximately 1893 through the 1920s. There is no mention of the company past the 1930s. Labeled left to right, the pedals are Mandolin, Orchestra, Expression, Soft, and Forte (Sustain). The Mandolin pedal produces a sound similar to a tremolo feel by bouncing a set of small hammers against the strings, enabling the piano to mimic a mandolin, guitar, banjo, zither and harp. The Orchestra (Orch) pedal used a similar approach, lowering a set of metal-tipped felt strips in between the hammers and the strings. This extended the life of the hammers when the Orch pedal was used, a good idea for practicing, and created an echo-like sound that mimicked playing in an orchestral hall.[16] [17]

The *pedalier* piano, or pedal piano, is a rare type of piano that includes a pedalboard so players can user their feet to play bass register notes, as on an organ. There are two types of pedal piano. On one, the pedal board is an integral part of the instrument, using the same strings and mechanism as the manual keyboard. The other, rarer type, consists of two independent pianos (each with separate mechanics and strings) placed one above the other—one for the hands and one for the feet. This was developed primarily as a practice instrument for organists, though there is a small repertoire written specifically for the instrument.

Construction

Many parts of a piano are made of materials selected for strength and longevity. This is especially true of the outer rim. It is most commonly made of hardwood, typically maple or beech, and its massiveness serves as an essentially immobile object from which the flexible soundboard can best vibrate. According to Harold A. Conklin,[18] the purpose of a sturdy rim is so that, "... the vibrational energy will stay as much as possible in the soundboard instead of dissipating uselessly in the case parts, which are inefficient radiators of sound."

Outer rim of Estonia grand piano during the manufacturing process

While many companies make hardwood rims are made by laminating thin (hence flexible) strips of hardwood plywood, bending them to the desired shape immediately after the application of glue, high end manufacturers like Bösendorfer use a solid piece of wood. Bösendorfer, the Austrian manufacturer of high quality pianos, constructs their rim from spruce, the very same wood that the soundboard is made from. Their idea is to concertedly involve the cabinet in the projection and coloration of sound. The *loss* of energy into the Bösendorfer case alters the instrument's tone, giving it perhaps less power but a complex and unusually resonant sound.

This view of the underside of a 182 cm (6 foot) grand piano shows, in order of distance from viewer: softwood braces, tapered soundboard ribs, soundboard. The metal rod at lower right is a humidity control device.

Theodore Steinway developed the bent plywood rim in 1880 to reduce manufacturing time and costs. The thick wooden posts on the underside (grands) or back (uprights) of the piano stabilize the rim structure, and are made of softwood for stability. The requirement of structural strength, fulfilled by stout hardwood and thick metal, makes a piano heavy. Even a small upright can weigh 136 kg (300 lb), and the Steinway concert grand (Model D) weighs 480 kg (990 lb). The largest piano built, the Fazioli F308, weighs 691 kg (1520 lb).

The pinblock, which holds the tuning pins in place, is another area where toughness is important. It is made of hardwood, (typically hard maple or beech), and is laminated for strength, stability and longevity. Piano strings (also called piano wire), which must endure years of extreme tension and hard blows, are made of high carbon steel. They are manufactured to vary as little as possible in diameter, since all deviations from uniformity introduce tonal distortion. The bass strings of a piano are made of a steel core wrapped with copper wire, to increase their mass whilst retaining flexibility. If all strings throughout the piano's compass were individual (monochord), the massive bass strings would overpower the upper ranges. Makers compensate for this with the use of double (bichord) strings in the tenor and triple (trichord) strings throughout the treble.

The plate (harp), or metal frame, of a piano is usually made of cast iron. A massive plate is advantageous. Since the strings vibrate from the plate at both ends, an insufficiently massive plate would absorb too much of the vibrational energy that should go to through the bridge to the soundboard. While some manufacturers use cast steel in their plates, most prefer cast iron. Cast iron is easy to cast and machine, has flexibility sufficient for piano use, is much more resistant to deformation than steel, and is especially tolerant of compression. Plate casting is an art, since dimensions are crucial and the iron shrinks about one percent during cooling.

Cast iron plate of a grand piano

Including an extremely large piece of metal in a piano is potentially an aesthetic handicap. Piano makers overcome this by polishing, painting, and decorating the plate. Plates often include the manufacturer's ornamental medallion. In an effort to make pianos lighter, Alcoa worked with Winter and Company piano manufacturers to make pianos using an aluminum plate during the 1940s. Aluminum piano plates were not widely accepted, and were discontinued.

The numerous grand parts and upright parts of a piano action are generally hardwood, e.g., maple, beech, or hornbeam. However, since WWII, makers have used some plastics. Early plastics were incorporated into some pianos in the late 1940s and 1950s, but proved disastrous because they lost strength after a few decades of use. Starting in 1961, the New York branch of the Steinway firm incorporated Teflon, a synthetic material developed by DuPont, for some parts of its *Permafree* grand action in place of cloth bushings, but abandoned the experiment in 1982 due to excessive friction and a "clicking" that developed over time. Teflon is "humidity stable" whereas the wood adjacent to the Teflon swells and shrinks with humidity changes, causing problems. More recently, the Kawai firm built pianos with action parts made of more modern materials such as carbon fiber reinforced plastic, and the piano parts manufacturer Wessell, Nickel and Gross has launched a new line of carefully engineered composite parts. Thus far these parts have performed reasonably, but it will take decades to know if they equal the longevity of wood.

Strings of a grand piano

In all but the poorest pianos the soundboard is made of solid spruce (that is, spruce boards glued together along the side grain). Spruce's high ratio of strength to weight minimizes acoustic impedance while offering strength sufficient to withstand the downward force of the strings. The best piano makers use quarter-sawn, defect-free spruce of close annular grain, carefully seasoning it over a long period before fabricating the soundboards. This is the identical material that is used in quality acoustic guitar soundboards. Cheap pianos often have plywood soundboards.

In the early years of piano construction, keys were commonly made from sugar pine. Today they are usually made of spruce or basswood. Spruce is typically used in high-quality pianos. Black keys were traditionally made of ebony, and the white keys were covered with strips of ivory. However, since ivory-yielding species are now endangered and protected by treaty, makers use plastics almost exclusively. Also, ivory tends to chip more easily than plastic. Legal ivory can still be obtained in limited quantities. The Yamaha firm invented a plastic called "Ivorite" that they claim mimics the look and feel of ivory. It has since been imitated by other makers.

Care and maintenance

The piano at the social center in the 19th century (Moritz von Schwind, 1868). The man at the piano is Franz Schubert.

Pianos need regular tuning to keep them on pitch, which is usually the internationally recognized standard concert pitch of A4 = 440 Hz. The hammers of pianos are voiced to compensate for gradual hardening, and other parts also need periodic regulation. Aged and worn pianos can be rebuilt or reconditioned. Often, by replacing a great number of their parts, they can perform as well as new pianos. Older pianos are often more settled and produce a warmer tone.

Piano moving should be done by trained piano movers using adequate manpower and the correct equipment for any particular piano's size and weight. Pianos are heavy yet delicate instruments. Over the years, professional piano movers have developed special techniques for transporting both grands and uprights, which prevent damage to the case and to the piano's mechanics.

Tuning

A piano tuner

The relationship between two pitches, called an interval, is the ratio of their absolute frequencies. Two different intervals are perceived as the same when the pairs of pitches involved share the same frequency ratio. The easiest intervals to identify, and the easiest intervals to tune, are those that are just, meaning they have a simple whole-number ratio. The term *temperament* refers to a tuning system that tempers the just intervals (usually the perfect fifth, which has the ratio 3:2) to satisfy another mathematical property; in equal temperament, a fifth is tempered by narrowing it slightly, achieved by flattening its upper pitch slightly, or raising its lower pitch slightly. A temperament system is also known as a set of **bearings**.

Tempering an interval causes it to beat, which is a fluctuation in perceived sound intensity due to interference between close (but unequal) pitches. The rate of beating is equal to the frequency differences of any harmonics that are present for both pitches and that coincide or nearly coincide.

Piano tuning is the act of adjusting the tensions of the piano's strings, thereby aligning the intervals among their tones so that the instrument is in tune. The meaning of the term *in tune* in the context of piano tuning is not simply a particular fixed set of pitches. Fine piano tuning carefully assesses the interaction among all notes of the chromatic scale, different for every piano, and thus requires slightly different pitches from any theoretical standard. Pianos are usually tuned to a modified version of the system called equal temperament (*see Piano key frequencies for the theoretical piano tuning*). In all systems of tuning, each pitch is derived from its relationship to a chosen fixed pitch, usually A440.

Role

The piano is a crucial instrument in Western classical music, jazz, film, television, and most other complex western musical genres. A large number of composers are proficient pianists—and because the piano keyboard offers an easy means of complex melodic and harmonic interplay—the piano is often used as a tool for composition.

See also

General	Technical	Related instruments	Other
• Jazz piano • Piano extended technique • Piano transcription • Piano trio • PianoForte Foundation • Street piano • String piano	• Agraffe • Aliquot stringing • Piano acoustics	• Digital piano • Electric piano • Electronic keyboard • Electronic piano • Harmonichord • Keyboard instruments • Keytar • Organ • Pipe organ • Player piano	• Chiroplast • Pianist • Pianos • List of classical pianists • List of films about pianists • List of piano brand names • List of piano makers

Notes

- *Encyclopædia Britannica* (available online by subscription) also includes much information on the piano. In the 1988 edition, the primary article can be found in "Musical Instruments".
- *Giraffes, black dragons, and other pianos: a technological history from Cristofori to the modern concert grand* by Edwin M. Good (1982, second ed., 2001, Stanford, Calif.: Stanford University Press) is a standard reference on the history of the piano.
- *The Early Pianoforte* by Stewart Pollens (1995, Cambridge: Cambridge University Press) is an authoritative work covering the ancestry of the piano, its invention by Cristofori, and the early stages of its subsequent evolution.
- *The New Grove Dictionary of Music and Musicians* (available online by subscription), contains a wealth of information. Main article: "Pianoforte".
- *The Piano Book* by Larry Fine (4th ed. Jamaica Plain, Massachusetts: Brookside Press, 2001; ISBN 1-929145-01-2) gives the basics of how pianos work, and a thorough evaluative survey of current pianos and their manufacturers. It also includes advice on buying and owning pianos.

References

[2] Percy A. Scholes (1975), *Oxford Companion to Music*, tenth ed., Oxford and New York: Oxford University Press ISBN 0 19 311306 6

[3] David R. Peterson (1994), "Acoustics of the hammered dulcimer, its history, and recent developments", *Journal of the Acoustical Society of America* **95** (5), p. 3002.

[4] Chapter 1.

[6] http://www.metmuseum.org/toah/hd/cris/hd_cris.htm

[7] . "Instrument: piano et forte genandt" [was] an expression Bach also used when acting as Silbermann's agent in 1749."

[8] http://mediatheque.cite-musique.fr/ClientBookLineCIMU/recherche/NoticeDetailleByID.asp?ID=0162147

[10] Macaulay, David. *The New How Things Work. From Levers to Lasers, Windmills to Web Sites, A Visual guide to the World of Machines.* Houghton Mifflin Company, United States. 1998. ISBN 0-395-93847-3. pp. 26–27.

[11] Physics of the Piano by the Piano Tuners Guild (http://www.physics.odu.edu/~hyde/Teaching/Phys332_Wk13.ppt)

[12] Reblitz, Arthur A. *Piano Servicing, Tuning, and Rebuilding. For the Professional, the Student, and the Hobbyist.* Vestal Press, Lanham Maryland. 1993. ISBN 1-879511-03-7 Pp. 203–215.

[13] Reblitz, Arthur A. *Piano Servicing, Tuning, and Rebuilding. For the Professional, the student, and the Hobbyist.* Vestal Press, Lanham Maryland. 1993. ISBN 1-879511-03-7 Pp. 203–215.

[14] "Wireless Piano Exhibited in Germany." (http://books.google.com/books?id=Nd8DAAAAMBAJ&pg=PA115&dq=1954+Popular+Mechanics+January&hl=en&sa=X&ei=Q3YzT6TaKu2o0AHgjvW_Ag&ved=0CDMQ6AEwATgK#v=onepage&q&f=true) *Popular Mechanics*, April 1954, p. 115, bottom of page.

Further reading

- (Dutch)

External links

- History of the Piano Forte (http://www.uk-piano.org/history/history.html), Association of Blind Piano Tuners, UK
- Section **Table of Music Pitches** of the Virginia Tech Multimedia Music Dictionary (http://www.music.vt.edu/musicdictionary/appendix/pitch/pitch.html)
- The Frederick Historical Piano Collection (http://www.frederickcollection.org/collection.html)
- The Pianofortes of Bartolomeo Cristofori, Heilbrunn Timeline of Art History, The Metropolitan Museum of Art (http://www.metmuseum.org/toah/hd/cris/hd_cris.htm)

Francis_Poulenc

Francis Jean Marcel Poulenc (French: [fʁɑ̃sis ʒɑ̃ maʁsɛl pulɛ̃k]; 7 January 1899 – 30 January 1963) was a French composer and a member of the French group Les Six. He composed art song, solo piano music, chamber music, oratorio, choral music, opera, ballet music, and orchestral music.

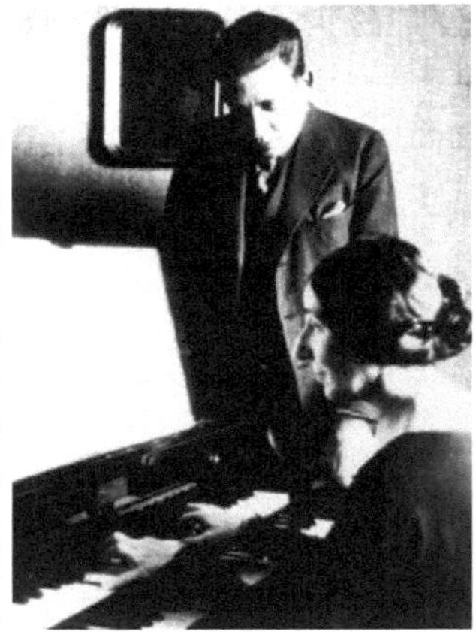

Poulenc and harpsichordist Wanda Landowska

Early life and education

Poulenc was born in Paris in 1899.[1] His father Émile Poulenc was a second generation director of the Poulenc, and later Rhône-Poulenc, chemical corporation.[2] His mother, an amateur pianist, taught him to play. In 1914, he was introduced to the Spanish pianist Ricardo Viñes, a champion of the music of Claude Debussy and Maurice Ravel. Poulenc became his pupil shortly afterwards,[3] and developed into a capable pianist[4] whose early compositions were dominated by the keyboard.

In 1916 a childhood friend, Raymonde Linossier (1897–1930), introduced Poulenc to Adrienne Monnier's bookshop, the *Maison des Amis des Livres*.[5] There he met avant-garde poets such as Guillaume Apollinaire, Max Jacob, Paul Éluard and Louis Aragon.[6] He later set many of their poems to music.

Career

Early works and Les Six

His first surviving composition, *Rapsodie Nègre* (1917), caught the attention of the Russian composer Igor Stravinsky.[7] Stravinsky was later instrumental in having the work published in London.[7]

Le bestiaire, ou Le cortège d'Orphée (1917) is a cycle of *mélodies* on poems by Guillaume Apollinaire. In 1918 Poulenc performed the premiere of his Sonata for Piano Four Hands with a fellow Viñes pupil, Marcelle Meyer.

Poulenc gave first performances of several of his new pieces: the Sonata for Two Clarinets, the Sonata for Piano Four Hands, a Sonata for Violin and Piano, and *Trois mouvements perpétuels,* at a series of concerts held from 1917 to 1920 in the studio of the painter Émile Lejeune in Montparnasse.[8] There Poulenc met other young composers, and together they formed Erik Satie's *Les nouveaux jeunes,* followed by Jean Cocteau's Les Six, a loose-knit group of young French and French-Swiss composers. Poulenc composed his *Valse en ut* for *L'Album des Six* (1920). He contributed to *Les mariés de la tour Eiffel* (1921) with *Discours du Général* and *La Baigneuse de Trouville.*

During the 1920s, Poulenc's most immediate influences were Chabrier, Debussy, Satie, and Stravinsky. He generally followed the irreverent, flippant aesthetic stance of Les Six with melodies influenced by Parisian music halls.

Les biches and first major works

Between 1921 and 1925, Poulenc received his first formal training in composition when he studied with Charles Koechlin (1921–25).[9] He remained a mostly self-taught composer.

Sergei Diaghilev commissioned ballet music from Poulenc for *Les biches*. This was performed by the Ballets Russes in January 1924, with settings by Marie Laurencin.

In 1926 Poulenc met the baritone Pierre Bernac, who became a very close friend. A great many of the chansons and melodies Poulenc wrote were composed for Bernac. They gave recitals throughout the world together from 1935 until 1959. In 1927, Poulenc bought *Le grand coteau*, a house close to Noizay in the Touraine, where he enjoyed the quiet atmosphere he needed to work.

In 1928 he composed the *Concert champêtre*, a piece for harpsichord and orchestra commissioned by Wanda Landowska. It was dedicated to Poulenc's lover, the painter Richard Chanlaire. Also in 1928, Poulenc recorded his *Trois mouvements perpétuels* and the *Trio for piano, oboe and bassoon*, and then *Le bestiaire*. He also started publishing musical reviews in *Les arts phoniques*.

He first performed his Concerto for two pianos and orchestra with Jacques Février in 1932. Also in 1932 *Le Bal Masqué*, a lighter piece, was performed privately at the Noailles.

The death of several friends and the composer Pierre-Octave Ferroud, followed by a pilgrimage to Rocamadour in 1935, led to a religious reawakening in 1936. Poulenc continued to compose light music, such as the *Quatre chansons pour enfants* (1934) on texts by Jean Nohain(fr) (1934). Some of his works were more sombre and austere. This is exemplified by his first sacred pieces, *Litanies à la vierge noire* ("Litanies to the Black Madonna") (1936) and the *Mass in G* (1937). The trend toward "new dimensions and greater depth" in the composer's style was continued by the song cycle *Tel jour, telle nuit* (1937) and *Concerto in G minor* for organ, strings, and timpani (1938).

World War II

During World War II, Poulenc, Durey and Auric joined the "Comité de Front National des Musiciens" created at the instigation of the French Communist Party in May 1941, led by Elsa Barraine and Roger Désormière.

One of his most popular songs is "Les chemins de l'amour". It was originally written as part of the incidental music for Jean Anouilh's play *Léocadia* (1940), but it achieved fame outside that context. The remainder of the *Léocadia* music is lost.[10]

In his humorous ballet, *Les animaux modèles* (1942, with Serge Lifar), Poulenc used the theme of a French patriotic song without the German officers noticing. In 1943, he set to music *Deux poèmes de Louis Aragon*, although the poet was a communist and a member of the French Resistance. This piece was given at the salle Gaveau in December 1943. The cantata *Figure humaine*(fr) was finished in 1943, but it was not publicly performed until 1945, in London. It set to music poems by Paul Éluard, notably *Liberté*, thousands of copies of which had been dropped over French territory by the Royal Air Force in 1942. Poulenc dedicated this cantata to his companion, Raymond Destouches (a driver who died in 1988). It is also in 1945 that Poulenc composed *L'histoire de Babar, le petit éléphant* (The Story of Babar, based on a children's book).

The post-war years

Although the score of his only opera bouffe, *Les mamelles de Tirésias* ("Tiresias' Breasts", after Guillaume Apollinaire's surrealist drama) was completed in 1945, it was not performed until 1947, after Poulenc met soprano Denise Duval. She also sang the roles of Blanche in *Dialogues of the Carmelites* (1956) and Elle in *La voix humaine* (1958). The song cycle *La courte paille* was written for her (and her child) in 1960, and *La dame de Monte Carlo* in 1961.

Poulenc's compositions continued, including chansons accompanied on the piano, choral music, secular (*Huit chansons françaises*, 1945) and religious (*Stabat Mater*, 1950) works, orchestral pieces (*Sinfonietta*, 1947), chamber music (particularly for wind instruments), works for one or two pianos such as *L'embarquement pour Cythère*, a valse musette (1951). He also composed film music, including *La Duchesse de Langeais* (1942) and Jean Anouilh's *Le voyageur sans bagage*.

Between 1947 and 1949, he became a radio host for *À bâtons rompus*. His choice of music was very diverse, including many pieces from the French repertoire.

He gave the first of his many concerts in the United States in 1948 with Pierre Bernac. During his time in the US, he met the American soprano Leontyne Price, who sang his chansons, and the composer Samuel Barber. His *Mélodies passagères* were performed in Paris in February 1952 by their dedicatees, Bernac and Poulenc.

In 1953, Poulenc started working on what was to become the opera *Dialogues of the Carmelites* based on a story by Georges Bernanos. Poulenc soon became obsessed with this work. He likened the sad fate of Blanche de la Force to the long agony of his boyfriend Lucien Roubert, a travelling salesman (1909–55) who died on the same day as Poulenc finished the piano version of the *Dialogues*. Poulenc adapted Bernanos' text for the libretto. The opera was first performed at La Scala in Italian, in January 1957 with Virginia Zeani singing the principal soprano role of Blanche. In June 1957, it was produced at the Paris Opera with Denise Duval as Blanche and Régine Crespin as Madame Lidoine. In September of that year, it was produced in the USA with Leontyne Price as Mme Lidoine; this was her first stage opera.

He followed the opera with *La voix humaine* (1958), a lyric tragedy based on Jean Cocteau's play. It was dedicated to Poulenc's last lover, Louis Gautier (a manual worker he met in 1957); they stayed together until Poulenc died.

Poulenc travelled to the US in 1960–61 for the American premieres of *Les mamelles de Tirésias* and *La voix humaine*. His *Gloria* for soprano solo, choir and orchestra was premiered in both Boston, conducted by Charles Munch, and in Paris, conducted by Georges Prêtre.

Poulenc published a book on Emmanuel Chabrier in 1961. He composed *Sept répons des ténèbres* in 1962. His last two works were premiered posthumously, in April and June 1963: the *Sonata for Oboe and Piano* was given by Pierre Pierlot and Jacques Février and the *Sonata for Clarinet and Piano* by Benny Goodman and Leonard Bernstein. Poulenc was particularly fond of woodwind instruments: having planned a set of sonatas for all of them, he lived to complete only four, for flute, oboe, clarinet, and the *Elégie* for French horn.

Poulenc had only one piano student, Gabriel Tacchino; he has performed and recorded all his piano music.[11] Poulenc was the pianist in many of his recordings, including some of his own songs (with Pierre Bernac, recorded in 1947, and Rose Dercourt) and the *Concerto for Two Pianos* (recorded in May 1957). He supervised the 1961 world premiere recording of his *Gloria,* which was conducted by Georges Prêtre. His recordings were released by RCA Victor and EMI.

Aesthetic and style

Critic Claude Rostand(fr) underlined the coexistence or alternation in Poulenc of the Catholic faith with artistic freedom. He wrote in a July 1950 *Paris-Presse* article that Poulenc was "A lover of life, mischievous, 'bon enfant,' tender and impertinent, melancholy and serenely mystical, half monk and half delinquent [*le moine et le voyou*]."[12] When the second part of his *Gloria* caused a scandal, Poulenc said, "I was thinking when I composed it of these frescoes by Gozzoli with angels sticking out their tongues, and of those grave benedictines I saw once playing football."

Poulenc's music is fundamentally tonal, although he made use of harmonic innovations such as pandiatonicism and chromatically altered chords. In some of his last works he used 12-tone rows, but Poulenc never questioned the validity of traditional tonic-dominant harmony. Lyrical melody pervades his music and underlies his important contributions to vocal music, particularly French art song.

Personal life

Some writers consider Poulenc one of the first openly gay composers.[] His first serious relationship was with painter Richard Chanlaire, to whom he wrote on his *Concert champêtre* score: "You have changed my life, you are the sunshine of my thirty years, a reason for living and working".[13] He also once said, "You know that I am as sincere in my faith, without any messianic screamings, as I am in my Parisian sexuality."[14]

Poulenc lived at 5, rue de Médicis, Paris.

In 1923, Poulenc was "unable to do anything" for two days after the death from typhoid fever of twenty-year-old novelist Raymond Radiguet, a close friend of Jean Cocteau. However, two weeks later he made lewd jokes about a male ballet dancer to Sergei Diaghilev at the rehearsals of *Les Biches*.[15] Then in 1930 came the death of Raymonde Linossier, a woman to whom he had proposed a marriage of convenience in 1928 and had remained a "cherished friend" until her death.[16] In 1936, Poulenc was profoundly affected by the death of another composer, Pierre-Octave Ferroud. This led him to his first visit to the shrine of the Black Virgin of Rocamadour. Here, before the statue of the Madonna with a young child on her lap, Poulenc experienced a life-changing transformation. Thereafter, he produced a sizeable output of liturgical music or compositions based on religious themes.

In 1949, Poulenc experienced the death of a friend, the artist Christian Bérard, for whom he composed his *Stabat Mater* (1950).

During the completion of his opera *Dialogues des Carmelites*, Poulenc suffered severe depression due to the copyright to Georges Bernanos' libretto. This eventually drove him to nervous breakdown.[17] However, this is arguably the point at which his later style emerged; his descent into anxiety and depression and subsequent recovery can be seen as the source of the serenity and lyricism in much of his later work.[18]

Poulenc died of heart failure in Paris on 30 January 1963 and is buried at the Père Lachaise Cemetery.[19]

Works

See List of compositions by Francis Poulenc.

Books

Autobiographical

- Francis Poulenc, *Moi et mes amis*, confidences recueillies par Stéphane Audel. Paris: La Palatine Ligugé, 1963, 206pp.
- Francis Poulenc, *Journal de mes mélodies*. Cicero, 1993, 160pp.
- Francis Poulenc, *À bâtons rompus (écrits radiophoniques, Journal de vacances, Feuilles américaines)*, écrits édités par Lucie Kayas. Arles: Actes Sud, 1999.
- Francis Poulenc, *J'écris ce qui me chante*, textes et entretiens, réunis, présentés et annotés par Nicolas Southon, Fayard, 2011, 920 p.

Correspondence

- Francis Poulenc *Echo and Source. Selected Correspondence 1915–1963*, translated and edited by Sidney Buckland. London: Gollancz, 1991, 448pp.
- Francis Poulenc, *Correspondance 1910–1963*, éditée par Myriam Chimènes. Paris: Fayard, 1994, 1128pp. ISBN 978-2-213-03020-3

Monographs

In chronological order:

- Roy, Jean. *Francis Poulenc*. Paris: Seghers, 1964.
- Hell, Henri. *Francis Poulenc*. Paris: Fayard, 1978.
- Bernac, Pierre. *Francis Poulenc et ses mélodies*. Paris: Buchet-Chastel, 1978.
- Daniel, Keith W. *Francis Poulenc: His Artistic Development and Musical Style*. Ann Arbor: UMI Research Press, 1982. ISBN 978-0-8357-1284-2.
- Bloch, Francine. *Phonographie de Francis Poulenc*. Paris: Bibliothèque Nationale, 1984.
- Mellers, Wilfrid. *Francis Poulenc*. Oxford and New York: Oxford University Press, 1993. ISBN 0-19-816337-1.
- Renaud Machart, *Poulenc*. Paris: Seuil, 1995.
- Ivry, Benjamin. *Francis Poulenc*. London: Phaidon Press Limited, 1996. ISBN 0-7148-3503-X.
- Buckland, Sidney, and Myriam Chimènes, eds. *Poulenc: Music, Art and Literature*. Aldershot: Ashgate, 1999.
- Carl B. Schmidt, *Entrancing Muse: A Documented Biography of Francis Poulenc*. London: Pendragon Pr, 2001.
- Burton, Richard. *Francis Poulenc*. Absolute Press, 2002.
- Ramaut, Alban. *Francis Poulenc et la voix*. Lyon: Symétrie, 2005.
- Basinger, Simon. *Les Cahiers de Francis Poulenc*. Paris: Collectif de l'Association F.Poulenc, 2008.
- Lacombe, Hervé. *Francis Poulenc*. Paris: Fayard, 2013.

References

[1] *Baker's Biographical Dictionary of Twentieth-Century Classical Musicians*. New York: Schirmer Books, 1997.
[2] Schmidt, p.4
[3] Schmidt, p.19-20
[4] Myriam Chimènes: 'Poulenc, Francis', Grove Music Online ed. L. Macy (Accessed [25 December 2006]), (http://www.grovemusic.com)
[5] Schmidt, p.26
[6] Schmidt, p.27
[7] *L'Avant-scène opéra, no. 52, 1983, p. 7* (http://books.google.ca/books?id=Bb0IAQAAMAAJ&q="Rhapsodie+nÃ¨gre"+Strawinsky&dq="Rhapsodie+nÃ¨gre"+Strawinsky&hl=en&sa=X&ei=ZQJuT8X0FcX40gHYvbXSBg&ved=0CEkQ6AEwBA)
[8] Schmidt, p.55
[10] Classical Archives (http://www.classicalarchives.com/work/100023.html#tvf=tracks&tv=about)
[11] Bach Cantatas (http://www.bach-cantatas.com/Bio/Tacchino-Gabriel.htm)
[14] Aldrich, Robert and Wotherspoon, Gary (Eds.) (2001). *Who's Who in Contemporary Gay & Lesbian History: From World War II to the Present Day*. New York: Routledge. ISBN 0-415-22974-X.

External links

- Francis Poulenc 1899–1963, the official website (French and English version) (http://www.poulenc.fr/en/?Home)
- Francis Poulenc (http://www.findagrave.com/cgi-bin/fg.cgi?page=gr&GRid=7729) at *Find a Grave*
- Free scores by Francis Poulenc in the Choral Public Domain Library (ChoralWiki)
- Free scores by Francis Poulenc at the International Music Score Library Project

C_major

The **C major scale** (often just **C** or **key of C**) consists of the pitches C, D, E, F, G, A, and B. Its key signature has no flats/sharps.

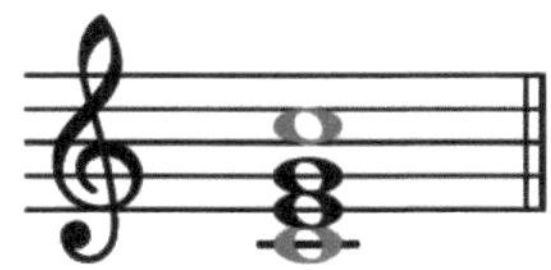

C major chord, root in red (Play). Note that the root is doubled at the octave.

Its relative minor is A minor, and its parallel minor is C minor.

C major is one of the most common key signatures used in music. Most transposing instruments playing in their home key are notated in C major; for example, a clarinet in B-flat sounding a B-flat major scale is notated as playing a C major scale. The white keys of the piano correspond to the C major scale. Among brass instruments, the more common trumpet is the trumpet in C, and the contra-bass tuba is in C. A pedal harp tuned to C major has all of its pedals in the middle position.

C major is often thought of as the simplest key, due to its lack of sharps or flats, and beginning piano students' first pieces are usually simple ones in this key; the first scales and arpeggios that students learn are also usually C major. However, going against this common practice, the Polish composer Frédéric Chopin regarded this scale as the most difficult to play with complete evenness, and he tended to give it last to his students. He regarded B major as the easiest scale to play on the piano, because the position of the black and white notes best fit the natural positions of the fingers, and so he often had students start with this scale. A C major scale lacks black keys and thus does not fit the natural positions of the fingers well.

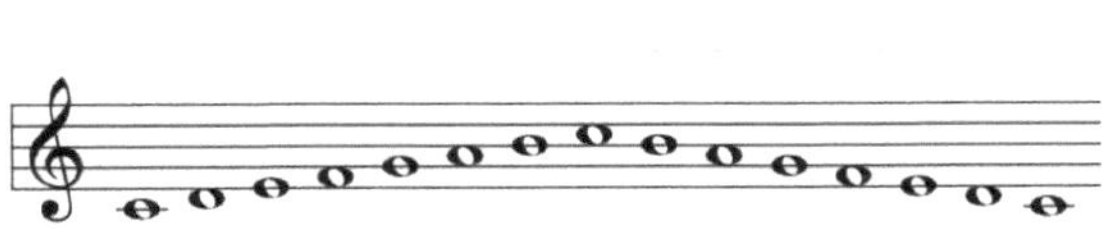

A one-octave **C major** scale.

Twenty of Joseph Haydn's 104 symphonies are in C major, making it his second most often used key, second only to D major. Of the 134 symphonies mistakenly attributed to Haydn that H. C. Robbins Landon lists in his catalog, 33 are in C major, more than any other key. Before the invention of the valve trumpet, Haydn did not write trumpet and timpani parts in his symphonies, except those in C major. H. C. Robbins Landon writes that it wasn't "until 1774 that Haydn uses trumpets and timpani in a key other than C major ... and then only sparingly." Most of Haydn's symphonies in C major are labelled "festive" and are of a primarily celebratory mood.[1] (See also List of symphonies in C major).

Many Masses and settings of *Te Deum* in the Classical era were in C major. Mozart wrote most of his Masses in C major and so did Haydn.[2]

Of Franz Schubert's two symphonies in the key, the first is nicknamed the "Little C major" and the second the "Great C major."

Many musicians have pointed out that every musical key conjures up specific feelings. American popular song writer Bob Dylan claimed the key of C major to "be the key of strength, but also the key of regret." "French composers such as Marc-Antoine Charpentier and Rameau generally thought of C major as a key for happy music, but Hector Berlioz in 1856 described it as "serious but deaf and dull." Ralph Vaughan Williams was impressed by Sibelius's Symphony No. 7 in C major and remarked that only Sibelius could make the key sound fresh. However, C major was a key of great importance in Sibelius's previous symphonies.[3] Claude Debussy, noted for composing music that avoided a particular key center, once said, "I do not believe in the supremacy of the C major scale."

In musical catalogs that sort the musical pieces by key, whether they go by semitones or along the circle of fifths, they almost always begin with those pieces in C major.

A notable modern use of the key is Terry Riley's *In C*.

Whereas traditionally key signatures were cancelled whenever the new key signature had fewer sharps or flats than the old key signature, in modern popular and commercial music, cancellation is only done when C major or A minor replaces another key.[4]

Well-known compositions in C major

- Johann Sebastian Bach
  - Toccata, Adagio and Fugue in C major, BWV 564
  - Cello Suite No. 3, BWV 1009
- Ludwig van Beethoven
  - Symphony No. 1, Op. 21
  - Piano Sonata No. 21, Op. 53 ("Waldstein")
  - Mass in C major
- Georges Bizet
  - Symphony in C
- Wolfgang Amadeus Mozart
  - Concerto for flute and harp, K. 299/297c
  - Piano Concerto No. 25, K. 503
  - Piano Sonata No. 16, K. 545
  - String Quartet No. 19, K. 465 ("Dissonance")
  - Symphony No. 36, K. 425 ("Linz")
  - Symphony No. 41, K. 551 ("Jupiter")
- Franz Schubert
  - *Wanderer Fantasy*, Op. 15 D. 760
  - Symphony No. 9, D. 944 ("Great")
  - String Quintet in C major, D. 956

- Robert Schumann
    - Fantasie in C, Op. 17
    - Symphony No. 2, Op. 61

References

[1] H. C. Robbins Landon, *The Symphonies of Joseph Haydn*. London: Universal Edition & Rockliff (1955): 227. "In the course of composing his first symphonies, the tonality of C major became indelibly impressed on Haydn's mind as the key of pomp, the key of C *alto* horns, trumpets and timpani, the vehicle for composing brillian and festive music, although at least during this period [the 1760s] he did not always reserve the tonality of C major for this particular kind of symphony: Nos. 2, 7 and 9, and possibly Nos. 25 and 30 ... are C major symphonies without the psychological manifestations inherent in most of the later works in this key. For the rest, however, the C major path is astonishingly clear; it can be traced from its inception, in Nos. 20, 32 and 37, through No. 33 and the more mature Nos. 38 and 41 to its synthesis in the *Maria Theresia* (No. 48) and No. 56. It continues with No. 50 and proceeds through Nos. 60, 63, 69, 82 and 90, reaching its final culmination in No. 97."

[2] James Webster & Georg Feder, *The New Grove Haydn*. New York: Macmillan (2002): 55. "The *Missa in tempora belli* ... in C features the bright, trumpet-dominated sound typical of masses in this key."

[3] Philip Coad, "Sibelius" in *A Guide to the Symphony* edited by Robert Layton. Oxford University Press. Sibelius's Seventh "is in C major, and a look back at the previous four symphonies [by Sibelius] will reveal how great the domination of C major has been [in his music]. It is the key of the Third, the relative major of the Fourth and the important 'neutral agent' in its Finale, the key which first forces away the tonic in the Fifth's Finale, and the principal opposition — the key of the brass — in the Sixth. Although it is now the tonic key, C major is also strongly associated with brass in the Seventh Symphony."

[4] Matthew Nicholl & Richard Grudzinski, *Music Notation: Preparing Scores and Parts*, ed. Jonathan Feist. Boston: Berklee Press (2007): 56. ISBN 978-0-87639-074-0. "In popular and commercial music, the old key signature is cancelled only if the new key is C major or A minor."

Further reading

- David Wyn Jones, "The Beginning of the Symphony", chapter in *A Guide to the Symphony* edited by Robert Layton. Oxford University Press.
- H. C. Robbins Landon, *Haydn: The Symphonies* BBC Music Guides

B-flat_minor

B♭ minor or **B-flat minor** is a minor scale based on B-flat, consisting of the pitches B♭, C, D♭, E♭, F, G♭, and A♭. Its key signature has five flats. The harmonic minor scale would use an A♮ instead of A♭.

B-flat natural minor scale ascending and descending. Play

B-flat harmonic minor scale ascending and descending. Play

B-flat melodic minor scale ascending and descending. Play

Its relative major is D-flat major, and its parallel major is B-flat major. Its enharmonic equivalent is A-sharp minor.

B-flat minor is traditionally a 'dark' key."[1] Important oboe solos in this key in the orchestral literature include the second movement of Tchaikovsky's Symphony No. 4, which depicts "the feeling that you get when you are all alone", in Tchaikovsky's words. Tchaikovsky's Piano Concerto No. 1 is also in B-flat minor. *An Alpine Symphony* by Richard Strauss begins and ends in B-flat minor.

The old valveless horn was barely capable of playing in B-flat minor; the only example found in 18th century music is a modulation that occurs in the first minuet of Franz Krommer's Concertino in D major, Op. 80.[2]

In classical music

- Samuel Barber's *Adagio for Strings*
- Frédéric Chopin's Piano Sonata No. 2, Nocturne Op. 9 No. 1, Scherzo No. 2 (ends in D-flat major)
- Sergei Rachmaninoff's Piano Sonata No. 2
- Dmitri Shostakovich's Symphony No. 13 and String Quartet No. 13
- Pyotr Ilyich Tchaikovsky's Piano Concerto No. 1, *Sérénade mélancolique* and *Marche Slave*
- Sir William Walton's Symphony No. 1

In popular music

All eight songs on The Raveonettes' *Whip It On* album are in the key of B-flat minor. Sune Rose Wagner said it was "Because Shostakovich's Symphony No. 13 was in B flat minor."[3]

British soul singer-songwriter Adele's 2007-2008 single, "Hometown Glory", is written in B♭ minor.[4]

The song "Say Say Say," written by Paul McCartney and Michael Jackson is written in B-flat minor.

The only chord used in the song "Papa Was a Rollin' Stone" by the Temptations is a B-flat minor chord.

"All of the Lights" by Kanye West is an example of a hip-hop song written in B-flat minor.

Little Talks by Indie band Of Monsters and Men is in the key of B-flat minor.[5]

B-flat minor is popular amongst saxophonists since B-flat is a common Saxophone tuning. Notable B-flat minor Saxophone solos in popular music include Gold by Spandau Ballet and Baker Street by Gerry Rafferty

References

[1] Wilfred Mellers, "Round and About in Górecki's Symphony No. 3" *Tempo* **168** 3 (1989): 23

[2] J. Murray Barbour, *Trumpets, Horns, and Music* (1964), p. 163

[4] Sheet Music Direct - Download Digital Sheet Music (http://sheetmusicdirect.us/search/productDetail.do?itemId=1000060614)

[5] Of Monsters And Men - Little Talks Sheet Music (Digital Download) (http://www.musicnotes.com/sheetmusic/mtdFPE.asp?ppn=MN0102852&ref=google)

E_minor

E minor (Em, Mim) is a minor scale based on the note E. The E natural minor scale ($\hat{1}$ $\hat{2}$ ♭$\hat{3}$ $\hat{4}$ $\hat{5}$ ♭$\hat{6}$ ♭$\hat{7}$) consists of the pitches E, F♯, G, A, B, C, and D. The E harmonic minor scale ($\hat{1}$ $\hat{2}$ ♭$\hat{3}$ $\hat{4}$ $\hat{5}$ ♭$\hat{6}$ $\hat{7}$) contains the natural 7, D♯, rather than the flatted 7, D – to align with the major dominant chord, B7 (B D♯ F♯ A).

E natural minor scale ascending and descending. Play

Its key signature has one sharp, F (*see below:* Scales and keys).

Its relative major is G major, and its parallel major is E major.

E harmonic minor scale ascending and descending. Play

Much of the classical guitar repertoire is in E minor, as this is a very natural key for the instrument. In standard tuning (E A D G B E), four of the instrument's six 'open' (unfretted) strings are part of the tonic chord. The key of E minor is also extremely popular in heavy metal music, as its tonic is the lowest note on a standard-tuned guitar.

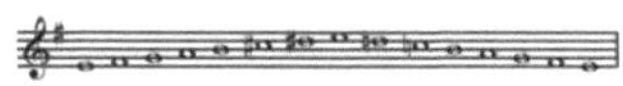

E melodic minor scale ascending and descending. Play

Well-known classical compositions in this key

- Symphony No. 4 – Johannes Brahms
- Nocturne in E minor – Frédéric Chopin
- Piano Concerto No. 1 – Frédéric Chopin
- Prelude in E minor – Frédéric Chopin
- Waltz in E minor - Frédéric Chopin
- Symphony No. 9 "From the New World" – Antonín Dvořák
- Slavonic Dance No. 2, Op. 72 – Antonín Dvořák
- Cello Concerto – Edward Elgar
- Symphony No. 44 "Trauer" – Joseph Haydn
- Violin Concerto in E minor – Felix Mendelssohn
- Caprice No. 3 – Niccolò Paganini
- Caprice No. 15 – Niccolò Paganini
- *Montagues and Capulets* - Sergei Prokofiev
- Symphony No. 2 – Sergei Rachmaninoff
- *Vocalise*, Op. 34, No. 14 – Sergei Rachmaninoff
- Symphony No. 10 – Dmitri Shostakovich
- Symphony No. 1 – Jean Sibelius
- Symphony No. 5 – Tchaikovsky
- Symphony No. 6 – Ralph Vaughan Williams

- Concerto for Bassoon, Strings, and continuo, RV 484 – Antonio Vivaldi
- Praeludium and Allegro – Fritz Kreisler

See also: list of symphonies in E minor

Manuel_de_Falla

Manuel de Falla y Matheu (23 November 1876 – 14 November 1946) was a Spanish composer. With Isaac Albéniz and Enrique Granados he is one of Spain's most important musicians of the first half of the 20th century. His image is on Spain's 1974 1000 Pesetas banknote.

Biography

Falla was born as Manuel María de los Dolores Falla y Matheu in Cádiz. He was the son of José María Falla y Franco and María Jesús Matheu y Zabal. His early teachers in music were his mother and grandfather; at the age of nine he was introduced to his first piano professor, Eloísa Galluzo. Little is known of that period of his life, but his relationship with his teacher was soon ended after she decided to then enter in a convent, Sisters of Charity, to become a nun. In 1889 he continued his piano lessons with Alejandro Odero and learned the techniques of harmony and counterpoint from Enrique Broca. At age 15 he became interested in literature and journalism and founded the literary magazines *El Burlón* and *El Cascabel*. In 1893 he was inspired by a concert of Edvard Grieg's works, later saying that at the time he felt that "my definitive vocation is music".

Madrid

In 1896 he moved to Madrid where he attended the Real Conservatorio de Música y Declamación. He studied piano with José Tragó, a colleague of Isaac Albéniz and composition with Felipe Pedrell. In 1897 he composed *Melodia* for cello and piano and dedicated it to Salvador Viniegra who hosted evenings of chamber music that Falla attended. In 1899, by unanimous vote, he was awarded the first prize at the piano competition at his school of music. He premiered his first works: *Romanza para violonchelo y piano*, *Nocturno para piano*, *Melodía para violonchelo y piano*, *Serenata andaluza para violín y piano*, and *Cuarteto en Sol y Mireya*. That same year he started to use *de* with his first surname, making *Manuel de Falla* the name he became known as from that time on. When only the surname is used, however, the *de* is omitted.

In 1900 he composed his *Canción para piano* and various other vocal and piano pieces. He premiered his *Serenata andaluza y Vals-Capricho para piano* in the Ateneo de Madrid. Due to the precarious financial position of his family he began to teach piano classes.

It was from Pedrell, during the Madrid period, that Falla became interested in native Andalusian music, particularly Andalusian flamenco (specifically *cante jondo*), the influence of which can be strongly felt in many of his works. Among his early pieces are a number of zarzuelas like *La Juana y la Petra* and *La casa de tócame Roque*. On 12

April 1902 he premiered *Los amores de la Inés* in the Teatro Cómico de Madrid. The same year he met the composer Joaquín Turina and saw his *Vals-Capricho* y *Serenata andaluza* published by the Society of Authors.

The following year he composed and performed *Allegro de concierto* (this was composed by Granados, not Falla)for the Madrid Royal Conservatory competition. Pianist Enrique Granados took first prize but the Society of Authors published Falla's works *Tus ojillos negros* and *Nocturno*. Falla then began his collaboration with composer Amadeo Vives on the zarzuelas *Prisionero de guerra*, *El cornetín de órdenes* and *La cruz de Malta* (only fragments of these works survive).

His first important work was the one-act opera *La vida breve* (*Life is Short*, or *The Brief Life*, written in 1905, though revised before its premiere in 1913). With a libretto by Carlos Fernández Shaw, *La vida breve* won Falla first prize in the Real Academia de Bellas Artes de San Fernando musical competition. In April 1905 he won the first prize in a piano competition sponsored by the firm of Ortiz and Cussó. On May 15 his work *Allegro de concierto* premiered in the Ateneo de Madrid and on November 13 the Real Academia presented him with his prize for *La vida breve*.

Paris

In 1907 at the advice of Joaquín Turina and Víctor Mirecki Larramat, Falla moved to Paris. There he met a number of composers who had an influence on his style, including the impressionists Maurice Ravel, Claude Debussy and Paul Dukas. In 1908 King Alfonso XIII awarded him a royal grant that enabled him to remain in Paris while he finished his *Cuatro piezas españolas*. Meanwhile, the dramatist Paul Milliet translated the libretto of *La vida breve* into French for its French premiere on 1 April 1913 at the Municipal Casino in Nice. In 1910 Falla met Igor Stravinsky and traveled briefly to London. He wrote *Siete canciones populares españolas* which he finished in mid-1914. Shortly after World War I began and Germany declared war, Falla returned to Madrid. While at no stage was he a prolific composer, it was then that he entered into his mature creative period.

Return to Madrid

In Madrid he composed several of his best known pieces, including:

- The nocturne for piano and orchestra *Nights in the Gardens of Spain* (*Noches en los jardines de España*, 1916)
- The ballet *El amor brujo* (*Wicked Love*, 1915) which includes the much excerpted and arranged *Ritual Fire Dance*
- The ballet *The Magistrate and the Miller's Wife* (*El corregidor y la molinera*) which, after revision, became *The Three-Cornered Hat* (*El sombrero de tres picos*, 1917) and was produced by Serge Diaghilev with set design and costumes by Pablo Picasso.

Granada period

From 1921 to 1939 Manuel de Falla lived in Granada, where he organized the Concurso de Cante Jondo in 1922. In Granada he wrote the puppet opera *El retablo de maese Pedro* (*Master Peter's Puppet Show*, 1923) and a concerto for harpsichord and chamber ensemble (*Harpsichord Concerto*, 1926). The puppet opera marked the first time the harpsichord had entered the modern orchestra; and the concerto was the first for harpsichord written in the 20th Century. Both of these works were written with Wanda Landowska in mind. In these works, the Spanish folk influence is somewhat less apparent than a kind of Stravinskian neoclassicism.

Also in Granada, Falla began work on the large-scale orchestral cantata *Atlántida* (*Atlantis*), based on the Catalan text *L'Atlántida* by Jacint Verdaguer. Falla considered *Atlántida* to be the most important of all his works; posterity has not agreed with this verdict, and performances of the piece have been extremely rare. Verdaguer's text gives a mythological account of how the submersion of Atlantis created the Atlantic ocean, thus separating Spain and Latin America, and how later the Spanish discovery of America reunited what had always belonged together.

In 1936, Falla tried but failed to prevent the murder of his close friend, the poet Federico García Lorca.

Argentina

Falla continued work on *Atlàntida* after moving to Argentina in 1939, following Francisco Franco's victory in the Spanish Civil War. The orchestration of the piece remained incomplete at his death and was completed posthumously by Ernesto Halffter. He also premiered his *Suite Homenajes* in Buenos Aires in November 1939. In 1940, he was named a Knight of the Order of King Alfonso X of Castile. Franco's government offered him a large pension if he would return to Spain, but he refused.

Falla did spend some time teaching in exile. Among his notable pupils was composer Rosa García Ascot. His health began to decline and he moved to a house in the mountains where he was tended by his sister María del Carmen de Falla. He died of cardiac arrest on 14 November 1946 in Alta Gracia, in the Argentine province of Córdoba. In 1947 his remains were brought back to Spain and entombed in the cathedral at Cádiz. One of the lasting honors to his memory is the Manuel de Falla Chair of Music in the Faculty of Philosophy and Letters at Complutense University of Madrid. His image appeared on Spanish currency notes for some years. Manuel de Falla never married and had no children.

Works

See **List of compositions by Manuel de Falla** and **List of works for the stage by Falla**.

References

- *Manuel de Falla and the Spanish Musical Renaissance* by Burnett James (Gollancz, London, 1979)
- *Manuel de Falla - a bibliography and research guide* by Andrew Budwig with Preface by Gilbert Chase (Garland Publishers, 1986)
- *Manuel de Falla* by Nancy Lee Harper (Greenwood Publishing Group, 1998)
- *Manuel De Falla and Modernism in Spain* by Carol A Hess (University of Chicago Press, 2001)
- *Falla* by Manuel Orozco Diaz (Barcelona: Salvat 1985)

See also

- John Brande Trend
- Spanish poetry

Recordings by de Falla

- Manuel de Falla 1876-1946 Grabaciones históricas; Centro de Documentación Musical de Andalucía. (Almaviva, HOM13080 [1]) (ref [2])
- *Rollos de Pianola (Obras de Albéniz, Granados, Turina, Ocón, Chapí, Alonso y Otros)* (Almaviva, DS - 0141 [3]) ASIN: B000GI34D6

External links

- The Manuel de Falla Foundation [4] in Spanish and English
- Free scores by Manuel de Falla at the International Music Score Library Project
- *Nana* by Manuel de Falla [5] from his *Siete Canciones Populares Españolas* (played by Duo Roldan)
- Arthur Rubinstein in De Falla's "Ritual Fire Dance" [6] on YouTube performed by Arthur Rubinstein
- Concert version for cello [7] performed by Julian Lloyd Webber (cello)
- Free scores by Manuel de Falla in the Choral Public Domain Library (ChoralWiki)

References

[1] http://www.almaviva.es/catalogodefault.php?disco=HOM13080&lang=en
[2] http://www.juntadeandalucia.es/cultura/centrodocumentacionmusical/opencms/publicaciones/2009/falla-1876-1946.html
[3] http://www.almaviva.es/catalogodefault.php?disco=DS%20-%200141&lang=en
[4] http://www.manueldefalla.com/
[5] http://www.iberoamericamusic.com/repertoire/nana-manuel-de-falla.html
[6] https://www.youtube.com/watch?v=Wj6_5qWZCDY
[7] http://www.youtube.com/watch?v=3pjjmgagF0E

Time_signature

<score> { \key c \major \time 3/4 \relative c *{ a f c \bar "*

Simple example of a time signature: here there are three (3) quarter-notes (4) per measure.

The **time signature** (also known as **meter signature**,[1] **metre signature**,[2] **bar signature**,[3] or **measure signature**[4]) is a notational convention used in Western musical notation to specify how many beats are in each bar and which note value constitutes one beat. In a musical score, the time signature appears at the beginning of the piece, as a time symbol or stacked numerals, such as 𝄴 or (read *common time* and *three four time*, respectively), immediately following the key signature or immediately following the clef symbol if the key signature is empty. A mid-score time signature, usually immediately following a barline, indicates a change of meter.

There are various types of time signatures, depending on whether the music follows simple rhythms or involves unusual shifting tempos, including: simple (such as or), compound (e.g., or), complex (e.g., or), mixed (e.g., & or or), additive (e.g.,), fractional (e.g.,), and irrational meters (e.g., or).

Simple time signatures

Basic time signatures: , also known as common time (𝄴); , also known as cut time or cut-common time (𝄵); plus ; ; and

Simple time signatures consist of two numerals, one stacked above the other:

- The *lower* numeral indicates the note value that represents one beat (the *beat unit*).
- The *upper* numeral indicates how many such beats there are in a bar.

For instance, means two quarter-note (crotchet) beats per bar— means three eighth-note (quaver) beats per bar.

The most common simple time signatures are , , and .

Notational variations in simple time

A semicircle, or 𝄴 , is sometimes used for time, also called common time or imperfect time. The symbol is derived from a broken circle used in music notation from the 14th through 16th centuries, where a full circle represented what today would be written in or time, and was called *tempus perfectum* (perfect time).[5] The symbol 𝄵 , a semicircle with a vertical line through is also a carry-over from the notational practice of late-Medieval and Renaissance music, where it signified *tempus imperfectum diminutum* (diminished imperfect time)—more precisely, a doubling of the speed, or *proportio dupla*, in duple meter.[6] In modern notation, it is used in place of and is called *alla breve* or, colloquially, *cut time* or *cut common time*.

Compound time signatures

In compound meter, subdivisions of the main beat (the upper number) split into three, not two, equal parts, so that a dotted note (half again longer than a regular note) becomes the beat unit. Compound time signatures are named as if they were simple time signatures, in which the one-third part of the beat unit is the beat, so the top number is commonly 6, 9 or 12 (multiples of 3). The lower number is most commonly an 8 (an eighth-note): as in or .

An example

is a simple signature that represents three quarter notes. It has a basic feel of (**Bold** denotes a stressed beat):

one *two three* (as in a waltz)

Each quarter note might comprise two eighth-notes (quavers) giving a total of six such notes, but it still retains that three-in-a-bar feel:

one and *two* and *three* and

: Theoretically, this can be thought of as the same as the six-quaver form of above with the only difference being that the eighth note is selected as the one-beat unit. But whereas the six quavers in had been in three groups of two, is practically understood to mean that they are in two groups of three, with a two-in-a-bar feel (**Bold** denotes a stressed beat):

one and a, **two** and a

Beat and time

Time signatures indicating two beats per bar (whether it is simple or compound) are called duple time; those with three beats to the bar are triple time. To the ear, a bar may seem like one singular beat. For example, a fast waltz, notated in time, may be described as being *one in a bar*. Terms such as *quadruple* (4), *quintuple* (5), and so on are also occasionally used.

Actual beat divisions

As mentioned above, though the score indicates a time, the actual beat division can be the whole bar, particularly at faster tempos. Correspondingly, at slow tempos the beat indicated by the time signature could in actual performance be divided into smaller units.

Interchangeability, rewriting meters

On a formal mathematical level the time signatures of, e.g., and are interchangeable. In a sense, *all* simple triple time signatures, such as , , , etc.—and all compound duple times, such as , and so on, are equivalent. A piece in can be easily rewritten in , simply by halving the length of the notes. Other time signature rewritings are possible: most commonly a simple time signature with triplets translates into a compound meter.

equals time at a different tempo Play

Though formally interchangeable, for a composer or performing musician, different time signatures often have different connotations. First, a smaller note value in the beat unit implies a more complex notation, which can affect ease of performance. Second, beaming affects the choice of actual beat divisions. It is, for example, more natural to use the quarter note/crotchet as a beat unit in or than the eight/quaver in or . Third, time signatures are traditionally associated with different music styles—it might seem strange to notate a rock tune in or .

equals time at a different tempo and requires the use of tuplets Play

Stress and meter

For all meters, the first beat (the downbeat, ignoring any anacrusis) is usually stressed (though not always, for example in reggae where the offbeats are stressed); in time signatures with four groups in the bar (such as and), the third beat is often also stressed, though to a lesser degree. This gives a regular pattern of stressed and unstressed beats, though notes on stressed beats are not necessarily louder or more important.

Most frequent time signatures

Simple time signatures		
(quadruple)	*Common time*: widely used in most forms of Western popular music. Most common time signature in rock, blues, country, funk, and pop[7]	 Simple quadruple drum pattern: divides four beats into two Play
(duple)	*Alla breve, cut time*: used for marches and fast orchestral music. Frequently occurs in musical theater. Sometimes called *in 2*, but may be notated in 4	 Simple quadruple drum pattern: divides four beats into two Play

(quadruple)	Never found in early music (which did not use numeric time signatures), and rare since 1600, though Brahms and other composers used it occasionally	
(duple)	Used for polkas or marches	 Simple duple drum pattern: divides two beats into two
(triple)	Used for waltzes, minuets, scherzi, country & western ballads, R&B, sometimes used in pop	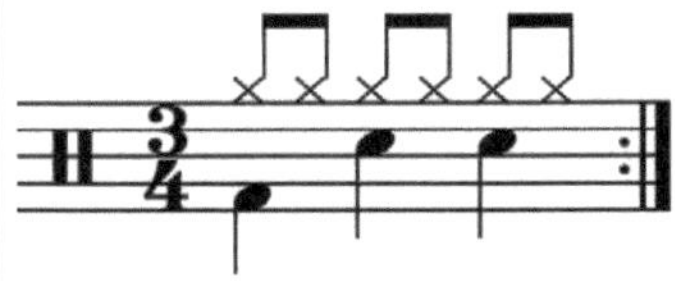 Simple triple drum pattern: divides three beats into two Play
(triple)	Also used for the above, but usually suggests higher tempo or shorter hypermeter	
	Compound time signatures	
(duple)	Double jigs, polkas, sega, salegy, tarantella, marches, barcarolles, Irish jigs, loures, and some rock music	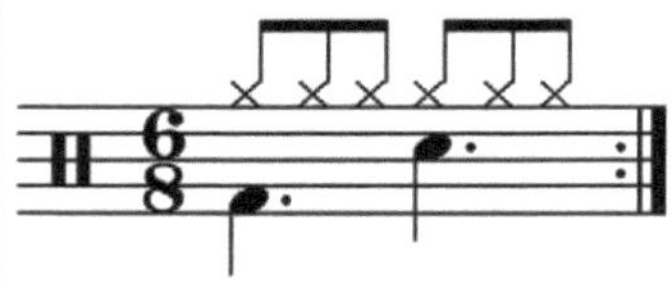 Compound duple drum pattern: divides two beats into three Play
(triple)	*Compound triple time*, used in triple ("slip") jigs, otherwise occurring rarely (*The Ride of the Valkyries*, Tchaikovsky's Fourth Symphony, and the final movement of the Bach Violin Concerto in A minor (BWV 1041)[8] are familiar examples. Debussy's *Clair de lune* and *Prélude à l'après-midi d'un faune* (opening bars) are in)	 Compound triple drum pattern: divides three beats into three Play
(quadruple)	Also common in slower blues (where it is called a *shuffle*) and doo-wop; also used more recently in rock music. Can also be heard in some jigs like *The Irish Washerwoman*. This is also the time signature of the Movement II *By the Brook* of Beethoven's *Symphony No 6 (the Pastoral)*	 Compound quadruple drum pattern: divides four beats into three Play

Video samples for the most frequent time signatures

For larger versions of the videos, click play, then go to More than About this file

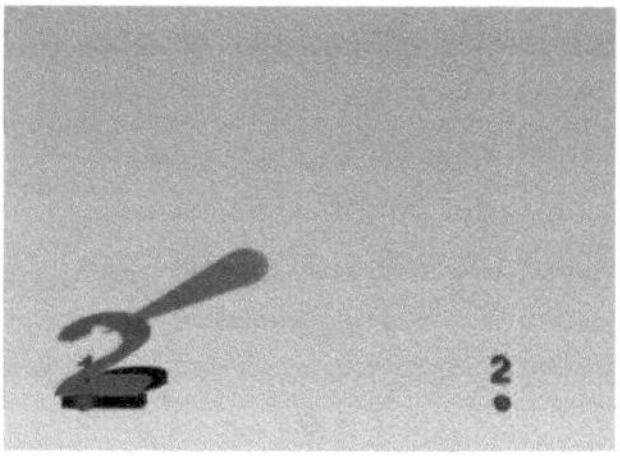 2/4 at a tempo of 60 bpm	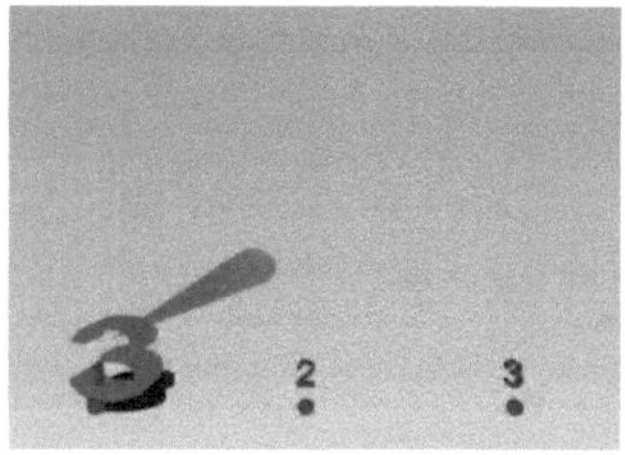 3/4 at a tempo of 60 bpm	 4/4 at a tempo of 60 bpm
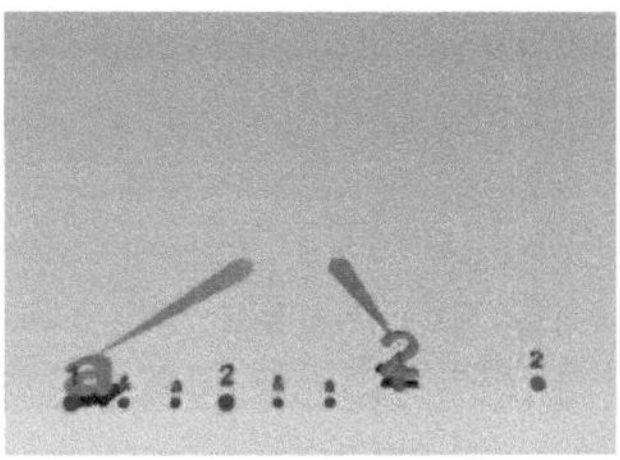 6/8 at tempo of 90 bpm	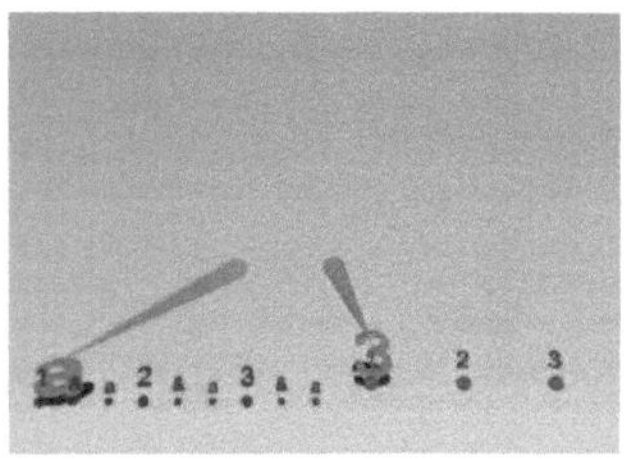 9/8 at tempo of 90 bpm	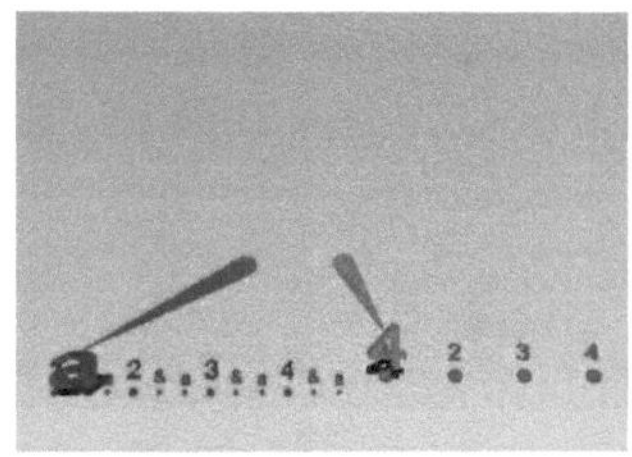 12/8 at tempo of 90 bpm

Complex time signatures

Signatures that do not fit the usual duple or triple categories are called *complex*, *asymmetric*, *irregular*, *unusual*, or *odd*—though these are broad terms, and usually a more specific description is appropriate. The term *odd meter*, however, sometimes describes time signatures in which the upper number is simply odd rather than even, including 3/4 and 9/8.[9] These more complex meters are common in some non-Western music, but rarely appeared in formal written Western music until the 19th century. The first deliberate quintuple meter pieces were apparently published in Spain between 1516 and 1520,[9] though other authorities reckon that the Delphic Hymns to Apollo (one by Athenaeus is entirely in quintuple meter, the other by Limenius predominantly so), carved on the exterior walls of the Athenian Treasury at Delphi in 128 BC, are probably earlier.[10] The third movement (Larghetto) of Chopin's Piano Sonata No. 1 (1828) is an early, but by no means the earliest, example of time in solo piano music. Reicha's Fugue 20 from his *Thirty-six Fugues*, published in 1803, is also for piano and is in . The waltz-like second movement of Tchaikovsky's *Pathétique* Symphony, often described as a *limping waltz*,[11] is a notable example of time in orchestral music. Examples from the 20th century include Holst's *Mars, the Bringer of War* and *Neptune, the Mystic* (both in) from the orchestral suite *The Planets*, Paul Hindemith's *Fugue Secunda in G*,() from *Ludus Tonalis*, the ending of Stravinsky's *Firebird* (), the fugue from Heitor Villa-Lobos's *Bachianas Brasileiras No. 9* () and the Mission Impossible theme by Lalo Schifrin (also in).

In the Western popular music tradition, unusual time signatures occur as well, with progressive rock in particular making frequent use of them. The use of shifting meters in The Beatles' *Strawberry Fields Forever* (1967) and the use of quintuple meter in their *Within You, Without You* (1967) are well-known examples,[12] as is Radiohead's

Paranoid Android (includes).[13]

Paul Desmond's jazz composition *Take Five*, in time, was one of a number of irregular-meter compositions that The Dave Brubeck Quartet played. They played other compositions in (*Eleven Four*), (*Unsquare Dance*)—and (*Blue Rondo à la Turk*), expressed as . This last is an example of a work in a signature that, despite appearing merely compound triple, is actually more complex.

However, such time signatures are only unusual in most Western music. Bulgarian dances, for example, use such meters extensively, including forms with 5, 7, 9, 11, 13, 15, 22, 25 and other numbers of beats per measure. These rhythms are notated as *additive rhythms* based on simple units, usually 2, 3 and 4 beats, though the notation fails to describe the metric "time bending" taking place, or compound meters. For example, the Bulgarian *Sedi Donka*, consists of 25 beats divided 7+7+11, where 7 is subdivided 3+2+2 and 11 is subdivided 2+2+3+2+2 or 4+3+4. See Variants below.

Video samples for complex time signatures

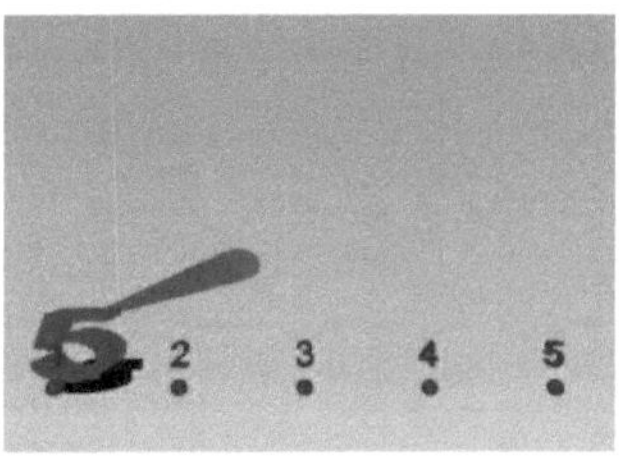

5/4 at 60 bpm

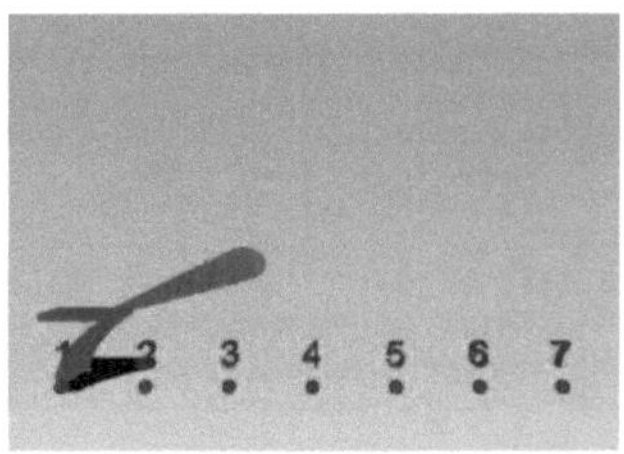

7/4 at 60 bpm

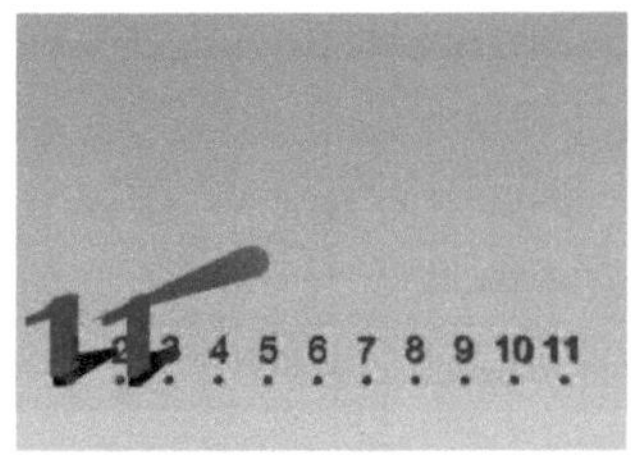

11/4 at 60 bpm

Rhythm of Blue Rondo à La Turk - consists of three measures of 2+2+2+3 followed by one measure of 3 + 3 + 3 and the cycle then repeats. Taking the smallest time unit as eighth notes, the arrows on the tempo dial show the tempi for ♪, ♩, ♩. and the measure beat. Starts slow, speeds up to usual tempo

Mixed meters

While time signatures usually express a regular pattern of beat stresses continuing through a piece (or at least a section), sometimes composers place a different time signature at the beginning of each bar, resulting in music with an extremely irregular rhythmic feel. In this case the time signatures are an aid to the performers, and not *necessarily* an indication of meter. The Promenade from Mussorgsky's *Pictures at an Exhibition* (1874) is a good example:

Mussorgsky's *Pictures at an Exhibition Promenade* Play

Igor Stravinsky's *The Rite of Spring* (1913) is famous for its "savage" rhythms:

In such cases, a convention that some composers follow (e.g., Olivier Messiaen, in his *La Nativité du Seigneur* and *Quatuor pour la fin du temps*) is to simply omit the time signature. Charles Ives's *Concord Sonata* has measure bars for select passages, but the majority of the work is unbarred.

Some pieces have no time signature, as there is no discernible meter. This is commonly known as free time. Sometimes one is provided (usually) so that the performer finds the piece easier to read, and simply has 'free time' written as a direction. Sometimes the word FREE is written downwards on the staff to indicate the piece is in free time. Erik Satie wrote many compositions that are ostensibly in free time, but actually follow an unstated and unchanging simple time signature. Later composers used this device effectively, writing music almost devoid of a discernibly regular pulse.

If two time signatures alternate repeatedly, sometimes the two signatures are placed together at the beginning of the piece or section, as in the chorus to the song *America* from *West Side Story*. In this case, the music alternates between (in two) in the first measure of each pair and (in three) in the second measure.

Video samples for mixed meters

Flamenco Bulerías with emphasis as **[12]** 1 2 **[3]** 4 5 **[6]** 7 **[8]** 9 **[10]** 11 - also the rhythm for the song America in West Side Story

Variants

Additive meters

To indicate more complex patterns of stresses, such as additive rhythms, more complex time signatures can be used. *Additive meters* have a pattern of beats that subdivide into smaller, irregular groups. Such meters are sometimes called *imperfect*, in contradistinction to *perfect meters*, in which the bar is first divided into equal units.[14]

For example, the signature

$$\frac{3+2+3}{8}$$

—which can be written (3+2+3)/8, means that there are 8 quaver beats in the bar, divided as the first of a group of three eighth notes (quavers) that are stressed, then the first of a group of two, then first of a group of three again. The stress pattern is usually counted as ***one***-two-three-**one**-two-**one**-two-three. This kind of time signature is commonly used to notate folk and non-Western types of music. In classical music, Béla Bartók and Olivier Messiaen have used such time signatures in their works. The first movement of Maurice Ravel's Piano Trio in A Minor is written in , in which the beats are likewise subdivided into 3 + 2 + 3 to reflect Basque dance rhythms.

Romanian musicologist Constantin Brăiloiu had a special interest in compound time signatures, developed while studying the traditional music of certain regions in his country. While investigating the origins of such unusual meters, he learned that they were even more characteristic of the traditional music of neighboring peoples (e.g., the Bulgarians). He suggested that such timings can be regarded as compounds of simple two-beat and three-beat meters, where an accent falls on every first beat, even though, for example in Bulgarian music, beat lengths of 1, 2, 3, 4 are used in the metric description. In addition, when focused only on stressed beats, simple time signatures can count as beats in a slower, compound time. However, there are two different-length beats in this resulting compound time, a one half-again longer than the short beat (or conversely, the short beat is $^2/_3$ the value of the long). This type of meter is called *aksak* (the Turkish word for "limping"), *impeded*, *jolting*, or *shaking*, and is described as an *irregular bichronic rhythm*. A certain amount of confusion for Western musicians is inevitable, since a measure they would likely regard as , for example, is a three-beat measure in *aksak*, with one long and two short beats (with subdivisions of 2+2+3, 2+3+2, or 3+2+2).[15]

Folk music may make use of metric time bends, so that the proportions of the performed metric beat time lengths differ from the exact proportions indicated by the metric. Depending on playing style of the same meter, the time bend can vary from non-existent to considerable; in the latter case, some musicologists may want to assign a different meter. For example, the Bulgarian tune Eleno Mome is written as 7=2+2+1+2, 13=4+4+2+3, 12=3+4+2+3, but an actual performance (e.g., Smithsonian Eleno Mome [16]) may be closer to 4+4+2+3.5. The Macedonian 3+2+2+3+2 meter is even more complicated, with heavier time bends, and use of quadruples on the threes. The metric beat time proportions may vary with the speed that the tune is played. The Swedish Boda Polska (Polska from the parish Boda) has a typical elongated second beat.

In Western classical music, metric time bend is used in the performance of the Viennese Waltz. Most Western music uses metric ratios of 2:1, 3:1, or 4:1 (two-, three- or four-beat time signatures)—in other words, integer ratios that make all beats equal in time length. So, relative to that, 3:2 and 4:3 ratios correspond to very distinctive metric rhythm profiles. Complex accentuation occurs in Western music, but as syncopation rather than as part of the metric accentuation.

Brăiloiu borrowed a term from Turkish medieval music theory: *aksak* (Turkish for crippled). Such compound time signatures fall under the "aksak rhythm" category that he introduced along with a couple more that should describe the rhythm figures in traditional music.[17] The term Brăiloiu revived had moderate success worldwide, but in Eastern Europe it is still frequently used. However, aksak rhythm figures occur not only in a few European countries, but on all continents, featuring various combinations of the two and three sequences. The longest are in Bulgaria. The shortest aksak rhythm figures follow the five-beat timing, comprising a two and a three (or three and two).

Video samples for additive meters

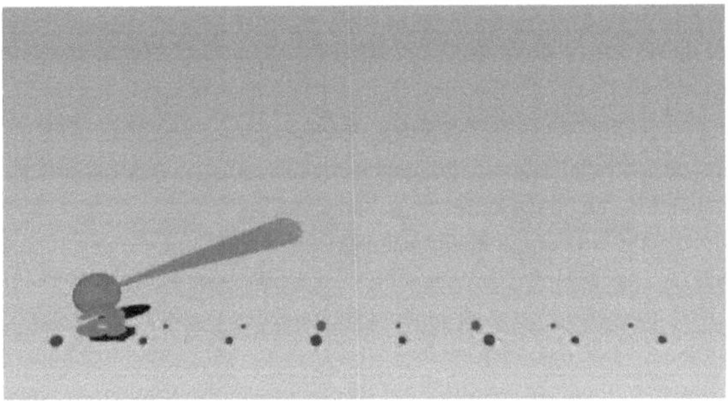

Time Signature 3 + 2 + 3 at 120 bpm

Other variants

Some composers have used fractional beats: for example, the time signature appears in Carlos Chávez's Piano Sonata No. 3 (1928) IV, m. 1.

Music educator Carl Orff proposed replacing the lower number of the time signature with an actual note image, as shown at right. This system eliminates the need for compound time signatures (described above), which are confusing to beginners. While this notation has not been adopted by music publishers generally (except in Orff's own compositions), it is used extensively in music education textbooks. Similarly, American composers George Crumb and Joseph Schwantner, among others, have used this system in many of their works.

Example of Orff's time signatures

Another possibility is to extend the barline where a time change is to take place above the top instrument's line in a score and to write the time signature there, and there only, saving the ink and effort that would have been spent writing it in each instrument's staff. Henryk Górecki's *Beatus Vir* is an example of this. Alternatively, music in a large score sometimes has time signatures written as very long, thin numbers covering the whole height of the score rather than replicating it on each staff; this is an aid to the conductor, who can see signature changes more easily.

Irrational meters

These are time signatures, used for so-called irrational bar lengths,[18] that have a denominator that is not a power of two (1, 2, 4, 8, 16, 32, etc.) (or, mathematically speaking, is not a dyadic rational). These are based on beats expressed in terms of fractions of full beats in the prevailing tempo—for example or .[18] For example, where implies a bar construction of four quarter-parts of a whole note (i.e., four quarter notes), implies a bar construction of four third-parts of it. These signatures are only of utility when juxtaposed with other signatures with varying denominators; a piece written entirely in , say, could be more legibly written out in .

Metric modulation is "a somewhat distant analogy".[18] It is arguable whether the use of these signatures makes metric relationships clearer or more obscure to the musician; it is always possible to write a passage using non-irrational signatures by specifying a relationship between some note length in the previous bar and some other in the succeeding one. Sometimes, successive metric relationships between bars are so convoluted that the pure use of irrational signatures would quickly render the notation extremely hard to penetrate. Good examples, written entirely in conventional signatures with the aid of between-bar specified metric relationships, occur a number of times in

John Adams' opera *Nixon in China* (1987), where the sole use of irrational signatures would quickly produce massive numerators and denominators.

Historically, this device has been prefigured wherever composers wrote tuplets. For example, a bar of 3 triplet crotchets could arguably be written as a bar of . Henry Cowell's piano piece *Fabric* (1920) employs separate divisions of the bar (anything from 1 to 9) for the three contrapuntal parts, using a scheme of shaped note heads to visually clarify the differences, but the pioneering of these signatures is largely due to Brian Ferneyhough, who says that he "find[s] that such 'irrational' measures serve as a useful buffer between local changes of event density and actual changes of base tempo.[18] Thomas Adès has also used them extensively—for example in *Traced Overhead* (1996), the second movement of which contains, among more conventional meters, bars in such signatures as , and .

A gradual process of diffusion into less rarefied musical circles seems underway. For example, John Pickard's *Eden*, commissioned for the 2005 finals of the National Brass Band Championships of Great Britain contains bars of .

Notationally, rather than using Cowell's elaborate series of notehead shapes, the same convention has been invoked as when normal tuplets are written; for example, one beat in is written as a normal quarter note, four quarter notes complete the bar, but the whole bar lasts only $^{4}/_{5}$ of a reference whole note, and a beat $^{1}/_{5}$ of one (or $^{4}/_{5}$ of a normal quarter note). This is notated in exactly the same way that one would write if one were writing the first four quarter notes of five quintuplet quarter notes.

This article uses *irrational* in the music theory sense, not the mathematical sense, where an irrational number is one that *cannot* be written as a ratio of whole numbers. However, at least one composition—Conlon Nancarrow's *Studies for Player Piano*—uses a time signature that is irrational in the mathematical sense. The piece contains a canon with a part augmented in the ratio $\sqrt{42}:1$ (approximately 6.48:1).

Video samples for irrational meters

These video samples show two time signatures combined to make a polymeter, since , say, in isolation, is identical to .

<table>
<tr>
<td>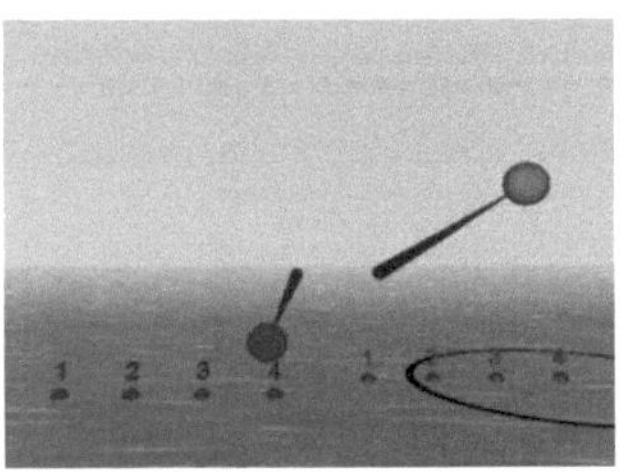
Polymeter ^{ 4 }_{ 4 } and ^{ 4 }_{ 3 } played together Has three beats of ^{ 4 }_{ 3 } to four beats of ^{ 4 }_{ 4 }</td>
<td>
Polymeter and played together
Has six beats of to four beats of</td>
<td>
Polymeter and played together
Has five beats of to three beats of . The displayed numbers count the underlying polyrhythm, which is 5:3</td>
</tr>
</table>

Early music usage

Mensural time signatures

In the 14th, 15th and 16th centuries, a period in which mensural notation was used, four basic *mensuration signs* determined the proportion between the two main units of rhythm. There were no measure or bar lines in music of this period; these signs, the ancestors of modern time signatures, indicate the ratio of duration between different note values. The relation between the breve and the semibreve was called **tempus**, and the relation between the semibreve and the minim was called **prolatio**. Unlike modern notation, the duration ratios between these different values was not always 2:1; it could be either 2:1 or 3:1, and that is what, amongst other things, these mensuration signs indicated. A ratio of 3:1 was called *complete*, perhaps a reference to the Trinity, and a ratio of 2:1 was called *incomplete*.

A circle used as a mensuration sign indicated *tempus perfectum* (a circle being a symbol of completeness), while an incomplete circle, resembling a letter C, indicated *tempus imperfectum.* Assuming the breve is a beat, this corresponds to the modern concepts of triple meter and duple meter, respectively. In either case, a dot in the center indicated *prolatio perfecta* while the absence of such a dot indicated *prolatio imperfecta*, corresponding to simple meter and compound meter.

A rough equivalence of these signs to modern meters would be:

- ⊙ corresponds to meter;
- ○ corresponds to meter;
- C· corresponds to meter;
- C corresponds to meter.

N.B.: in modern compound meters the beat is a dotted note value, such as a dotted quarter, because the ratios of the modern note value hierarchy are always 2:1. Dotted notes were never used in this way in the mensural period; the main beat unit was always a simple (undotted) note value.

Proportions

Another set of signs in mensural notation specified the metric proportions of one section to another, similar to a metric modulation. A few common signs are shown:[19]

- ¢ *tempus imperfectum diminutum*, 1:2 proportion (twice as fast);
- ⦶ *tempus perfectum diminutum*, 1:2 proportion (twice as fast);
- **C3** or just **3** *proportio tripla*, 1:3 proportion (three times as fast, similar to triplets).

Often the ratio was expressed as two numbers, one above the other,[20] looking similar to a modern time signature, though it could have values such as , which a conventional time signature could not.

Some proportional signs were not used consistently from one place or century to another. In addition, certain composers delighted in creating "puzzle" compositions that were intentionally difficult to decipher.

In particular, when the sign ¢ was encountered, the tactus (beat) changed from the usual semibreve to the breve, a circumstance called *alla breve*. This term has been sustained to the present day, and though now it means the beat is a minim (half note), in contradiction to the literal meaning of the phrase, it still indicates that the beat has changed to a longer note value.

See also

- Schaffel
- Tala

References

[1] Alexander R. Brinkman, *Pascal Programming for Music Research* (Chicago: University of Chicago Press, 1990): 443, 450–63, 757, 759, 767. ISBN 0226075079; Mary Elizabeth Clark and David Carr Glover, *Piano Theory: Primer Level* (Miami: Belwin Mills, 1967): 12; Steven M. Demorest, *Building Choral Excellence: Teaching Sight-Singing in the Choral Rehearsal* (Oxford and New York: Oxford University Press, 2003): 66. ISBN 0195165500; William Duckworth, *A Creative Approach to Music Fundamentals*, eleventh edition (Boston, MA: Schirmer Cengage Learning, 2013): 54, 59, 379. ISBN 0840029993; Edwin Gordon, *Tonal and Rhythm Patterns: An Objective Analysis: A Taxonomy of Tonal Patterns and Rhythm Patterns and Seminal Experimental Evidence of Their Difficulty and Growth Rate* (Albany: SUNY Press, 1976): 36, 37, 54, 55, 57. ISBN 0873953541; Demar Irvine, Reinhard G. Pauly, Mark A. Radice, *Irvine's Writing about Music*, third edition (Portland, Oregon: Amadeus Press, 1999): 209–210. ISBN 1574670492.

[2] Henry Cowell and David Nicholls, *New Musical Resources*, third edition (Cambridge and New York: Cambridge University Press, 1996): 63. ISBN 0521496519 (cloth); ISBN 0521499747 (pbk); Cynthia M. Gessele, "Thiéme, Frédéric [Thieme, Friedrich]", *The New Grove Dictionary of Music and Musicians*, second edition, edited by Stanley Sadie and John Tyrrell (London: Macmillan Publishers, 2001); James L. Zychowicz, *Mahler's Fourth Symphony* (Oxford and New York: Oxford University Press, 2005): 82–83, 107. ISBN 0195181654.

[3] Cuthbert Morton Girdlestone, *Jean-Philippe Rameau: His Life and Work*, revised and enlarged edition (New York: Dover Publications, 1969): 116. ISBN 0-486-21416-8.

[4] Edwin Gordon, *Rhythm: Contrasting the Implications of Audiation and Notation* (Chicago: GIA Publications, 2000): 111. ISBN 1579990983.

[6] Willi Apel, *The Notation of Polyphonic Music 900–1600*, fifth edition, revised and with commentary; The Medieval Academy of America Publication no. 38 (Cambridge, Massachusetts: The Medieval Academy of America, 1953): 147–48.

[7] Scott Schroedl, *Play Drums Today! A Complete Guide to the Basics: Level One* (Milwaukee: Hal Leonard Corporation, 2001), p. 42. ISBN 0-634-02185-0.

[8] *See* File:Bach BVW 1041 Allegro Assai.png for an excerpt from the violin part of the final movement.

[9] Tim Emmons, *Odd Meter Bass: Playing Odd Time Signatures Made Easy* (Van Nuys: Alfred Publishing, 2008): 4. ISBN 978-0-7390-4081-2. "What is an 'odd meter'?...A complete definition would begin with the idea of music organized in repeating rhythmic groups of three, five, seven, nine, eleven, thirteen, fifteen, etc."

[10] Egert Pöhlmann and Martin L. West, *Documents of Ancient Greek Music: The Extant Melodies and Fragments*, edited and transcribed with commentary by Egert Pöhlmann and Martin L. West (Oxford: Clarendon Press, 2001): 70–71 and 85. ISBN 0-19-815223-X.

[12] Edward Macan, *Rocking the Classics: English Progressive Rock and the Counterculture* (New York: Oxford University Press, 1997): 48. ISBN 978-0-19-509888-4.

[13] Radiohead (musical group). *OK Computer*, vocal score with guitar accompaniment and tablature (Essex, England: IMP International Music Publications; Miami, FL: Warner Bros. Publications; Van Nuys, Calif.: Alfred Music Co., Inc., 1997): . ISBN 0-7579-9166-1.

[14] Gardner Read, *Music Notation: A Manual of Modern Practice* (Boston: Allyn and Bacon, Inc., 1964): .

[15] Constantin Brăiloiu, *Le rythme Aksak*, *Revue de Musicologie* 33, nos. 99 and 100 (December 1951): 71–108. Citation on pp. 75–76.

[16] http://www.smithsonianglobalsound.org/searchresults.aspx?sPhrase=Eleno%20Mome&sType='phrase'

[17] Gheorghe Oprea, *Folclorul muzical românesc* (Bucharest: Ed. Muzicala, 2002), . ISBN 973-42-0304-5.

[18] "Brian Ferneyhough" (http://web.archive.org/web/20110721014850/http://www.sospeso.com/contents/articles/ferneyhough_p1.html), *The Ensemble Sospeso*

[19] Willi Apel, *The Notation of Polyphonic Music 900–1600*, fifth edition, revised with commentary; The Medieval Academy of America Publication no. 38 (Cambridge, Massachusetts: The Mediaeval Academy of America, 1953), p. 148.

[20] Willi Apel, *The Notation of Polyphonic Music 900–1600*, fifth edition, revised with commentary; The Medieval Academy of America Publication no. 38 (Cambridge, Massachusetts: The Mediaeval Academy of America, 1953), p. 147.

External links

- Grateful Dead songs with unusual time signatures (http://www3.clearlight.com/~acsa/rhythm.htm) (Grateful Dead)
- Odd Time Obsessed Internet Radio (http://www.oddtimeobsessed.com) - dedicated to "odd" meters
- More video samples of many time signatures (http://bouncemetronome.com/video-resources) - made with Bounce Metronome Pro (http://bouncemetronome.com) a program that can play all the time signatures mentioned in this article, even the ones that are irrational in the mathematical sense, like

El_amor_brujo

El amor brujo (*Love, the Magician*, rarely translated as *Wedded by Witchcraft*) is a piece of music originally composed by Manuel de Falla for a chamber group, then re-scored as a symphonic suite, and eventually as a ballet. The texts were by Gregorio Martínez Sierra.

The work is distinctively Andalusian in character with the songs in the Andalusian Spanish dialect of the Gypsies. The music contains moments of remarkable beauty and originality; it includes the celebrated *Ritual Fire Dance*, *Cancion del Fuego Fatuo* (Will-o'-the-Wisp) and the *Dance of Terror*.

Versions and performance history

El amor brujo was initially commissioned in 1914-15 as a *gitanería* (gypsy piece) by Pastora Imperio, a renowned flamenco gypsy dancer. It was scored for cantaora voice, actors, and chamber orchestra. It was performed at the Teatro Lara, Madrid on 15 April 1915 but was not successful.

The following year, Falla revised the work for symphonic orchestra, with three short songs for mezzo-soprano. This version was performed on 28 March 1916 with the Madrid Symphony Orchestra under Enrique Fernández Arbós.

In 1925, de Falla transformed it into a one-act 'ballet pantomímico'. The work's United States premiere was presented by the Philadelphia Civic Opera Company at Philadelphia's Metropolitan Opera House on March 17, 1927 with mezzo-soprano Kathryn Noll and conductor Alexander Smallens.

Synopsis

El amor brujo is the story of a young Andalusian gypsy girl called Candela. As a girl, she was promised to be married to another man (then a boy), although her affections are directed to Carmelo. After many years Candela's husband has died (at the hands of Lucía's husband) but he continues to haunt his wife.

The entire villageknows about the haunting but still brands Candela as crazy because she dances every night with him (Danza del terror). Candela, now a widow, is free to establish a relationship with Carmelo, but continues to being haunted by herhusband.

After a conversation with other women of the village, Candela finally comes to realise that her husband was unfaithful to her, despite all the efforts that she did to make their marriage work.

Candela and Carmelo get advice that a ritual dance is necessary to cast the ghost off (Danza ritual del fuego), but it does not work. The ghost is obsessed with Candela's soul.

Candela manages to trick Lucía to come that night, with the excuse of hooking her up with Carmelo. As she turns up, the nightly ritual of Candela's dance with her husband begins, but at the last moment Candela moves away from her husband and Lucía is taken away by her now dead lover (Danza del juego de amor).

Dawn breaks, Candela and Carmelo are now truly free to enjoy their love.

Movements

1. *Introducción y escena* ('Introduction and scene')
2. *En la cueva* ('In the cave')
3. *Canción del amor dolido* ('Song of suffering love')
4. *El aparecido (El espectro)* ('The apparition')
5. *Danza del terror* ('Dance of terror')
6. *El círculo mágico (Romance del pescador)* ('The magic circle')
7. *A media noche: los sortilegios*
8. *Danza ritual del fuego ('Ritual fire dance')*
9. *Escena* ('Scene')
10. *Canción del fuego fatuo* ('Song of the will-o'-the-wisp')
11. *Pantomima* ('Pantomime)
12. *Danza del juego de amor* ('Dance of the game of love')
13. *Final - las campanas del amanecer* ('Finale – the bells of sunrise')

Recordings

Chamber music version (1915)

- 1991: Josep Pons with the orchestra of the Teatre Lliure, 'cantaora' Ginesa Ortega. Harmonia Mundi HMC905213

Symphonic version

- 1946 (February 5): Fritz Reiner with the Pittsburgh Symphony Orchestra, Contralto Carol Brice. Columbia Masterworks MM-633 (3 12" 78 RPM discs); Also Columbia LP ML-2006.
- 1953: Ataúlfo Argenta with the Orchestre de la Société des Concerts du Conservatoire, mezzo-soprano Ana-Maria Iriarte. EMI 7243 5 69235 2 2
- 1955: Ernest Ansermet with the Orchestre de la Suisse Romande, mezzo-soprano Marina de Gabarain, Decca 417 691-2
- 1960: Leopold Stokowski with the Philadelphia Orchestra, mezzo-soprano Shirley Verrett-Carter, Columbia MS 6147
- 1966: Rafael Frühbeck de Burgos with the New Philharmonia Orchestra, mezzo-soprano Nati Mistral, Decca 417 786-2
- ? Carlo Maria Giulini with the Philharmonia Orchestra, soprano Victoria de los Ángeles EMI 7 69 037 2
- 1978 Luis Antonio Garcia Navarro with the London Symphony Orchestra, mezzo-soprano Teresa Berganza, Deutsche Grammophon 429181-2
- 1996 Edmon Colomer with the Orquestra Simfònica de Barcelona, 'cantaora' Esperanza Fernandez. Valois Auvidis V 4768.

Films

In 1967 Francisco Rovira Beleta directed a film version. It was nominated for the Academy Award for Best Foreign Language Film, but lost to Jiří Menzel's *Closely Observed Trains*. However it won the "National Syndicate of Spectacle, Spain" award.

In 1986, Spanish director Carlos Saura directed *El amor brujo* based on the ballet, starring, and choreographed by, Antonio Gades. It was the third in his trilogy of dance films, following *Bodas de sangre* (*Blood Wedding*) and *Carmen*. The film filled out the story with spoken dialogue, but nevertheless used the entire score of the ballet, along with additional songs and dances performed by characters in the film. The Orquesta Nacional de España was conducted by Jesús López-Cobos, and the cante jondo singer heard on the soundtrack was the late Rocío Jurado. A soundtrack album, now out of print, was issued by EMI.

Music

The section *"Cancion del Fuego Fatuo"* was covered by jazz musician Miles Davis on his album *Sketches of Spain (1959)* as *"Will O' the Wisp"*.

References

- Kennedy, Michael (2006), *The Oxford Dictionary of Music*, 985 pages, ISBN 0-19-861459-4

External links

- An essay on El amor brujo [1] by Peter Gutmann, accessed 25 December 2009
- El amor brujo [2] from the Martha Argerich Project
- YouTube: Ritual Fire Dance from *El amor brujo*, accessed 25 December 2009 [3]
- Libretto on Spanish Wikisource, accessed 25 December 2009 [4]
- Notes on El Amor Brujo by Sato Moughalian, Artistic Director, Perspectives Ensemble [5]

References

[1] http://www.classicalnotes.net/classics/falla.html

[2] http://www.rtsi.ch/trasm/argerich/welcome.cfm?lng=1&ids=493&idc=3947

[3] http://www.youtube.com/watch?v=Ftd8tIdiYq4&feature=related

[4] http://es.wikisource.org/wiki/El_amor_brujo_(libreto)

[5] http://www.perspectivesensemble.com/blog/perspectives-ensemble-notes/manuel-de-falla-el-amor-brujo

Article Sources and Contributors

Trois_novelettes_(Poulenc) *Source*: http://en.wikipedia.org/w/index.php?title=Trois_novelettes_%28Poulenc%29 *Contributors*: BD2412, DTOx, EdgeNavidad, Felix Folio Secundus, Iridescent, JackofOz, Midway, Rich Farmbrough, 1 anonymous edits

Novelette *Source*: http://en.wikipedia.org/w/index.php?title=Novelette *Contributors*: Aleksey Vladimirovich, Altenmann, Bongwarrior, Borislav, Chinanski, Darkpoet, DocWatson42, Dreamafter, Elphion, Evan Robidoux, Evb-wiki, Fritz freiheit, GSGSGSG, Ghirlandajo, Gjjm, Grandpallama, GregorB, Gregory of nyssa, Gymnophoria, Happysailor, Henry Merrivale, JamieJones, Jeff Silvers, Jfpierce, Jnestorius, Joelson0, JulesH, Jwy, Kayau, Kevinalewis, Kookyunii, Krash, Lamro, Lquilter, Markjoseph125, MasterRegal, Mike Klaassen, MikeRM, Mipadi, Oleg-ch, Owen, Ph89, PhilipC, ProveIt, Randwicked, Red banksy, ResAliens, Robertgreer, Sabiona, Sadads, Savant13, Scott197827, Smilo Don, Tenebrae, Tide rolls, Two lips, Verda stelo, Wicyclop2, Xanthoxyl, Zappa93, احمد غامدي24, 51 anonymous edits

Piano *Source*: http://en.wikipedia.org/w/index.php?title=Piano *Contributors*: (jarbarf), -Kerplunk-, 12345klfdsjkl, 1ForTheMoney, 1Si61515, 2001:558:6045:A0:391F:B005:179D:8DD9, 2001:7C0:E701:4C34:C0D6:A7EC:D0EE:B06A, 202.139.170.xxx, 23pokrzywa, 2601:B:9880:17:A43B:1FD4:B12A:8AF8, 28421u2232nfenfcenc, 2help, 336, 64.81.86.xxx, 7, 88tuner, 90 Auto, A Jewish Family, A Stop at Willoughby, A. Parrot, A3RO, A455bcd9, AAP inc., ABF, AEMoreira042281, ARUNKUMAR P.R, Abob1234567890, Abster241, Ac8720, Acather96, Accountforpicture, AceKingQueenJack, Acsenray, Adashiel, Addshore, AdjustShift, Adriaan, Agateller, Agsbphysics, Ahoerstemeier, Aitias, Ajcounter, Ajs15, AlanM1, Alansohn, Aldis90, Alegoo92, Alexius08, Alexthe5th, Alexwill23, Alibarbour, All Is One, Allstarecho, Alpha 4615, Alsandro, Altenmann, Altes2009, Alton, Amok116, Andonic, Andrewa, Andrewrp, Andy M. Wang, Andycjp, Angmering, Animum, Aniten21, Annie L, Anniesotm, Anonymous Dissident, Antandrus, Antonio Lopez, Aotake, Apparition11, Aravis222, Arbor to SJ, Arctic Night, Ardonik, ArglebargleIV, Arjun01, Asdfkidasdf, Ashandpikachu, Ashley thomas80, Asphatasawhale, Astroview120mm, Athanasius1, AtheWeatherman, AtomBoy, AtticusX, Augustus the Pony, Austin 88888, Autoerrant, Avb, Avoided, B-Dubbers, B137, BILLUS MACGEE, BShiplet, BUF4Life, Bachbust100, Badagnani, Badseed, Baghshy, Bahar101, Bald Zebra, Bantosh, Barrrakuda, Bart133, Baseball Watcher, Bb3bb3bb, Bdesham, Bdiscoe, BearGuard, Befeyalcin, Bella Swan, Bemoeial, Benbread, Bendzh, Benechanuk, Benjpianist, Benmon1, Betsys99, Bff711, Billyfisher, Billyfisher100, Billyfisher100100100, Billytrousers, Binksternet, Birri85, Blaxthos, Bleh fu, Blehfu, Blinggu, Blouis79, Blue520, BlueDevil, Bmusician, Bobo192, Boccobrock, Bocianski, Bogey97, Bollinger, Bomac, Bongwarrior, Bookermorgan, Borat134, Bovineboy2008, BrianWilloughby, Brianga, Bridgecross, Briguychau, Brissone, BritishWatcher, Broadwayfreak99, Brucejohn, Brumpz, Buchanae, Burntsauce, Bushcarrot, Buxtehude, Bxj, C.Fred, CL, CPMartin, CTF83!, CWii, Caggymorgan, Caltas, CalumH93, Calvin 1998, Calvinballing, Camembert, Cameronandboas7, Canadian-Bacon, CanadianLinuxUser, Canderson7, CanisRufus, CapitalR, Capricorn42, Carboncopy, Carchuleta, Carebearac3, Catgut, Cavarrone, Cb77305, Cedricsimsmahrlig, Cfortunato, Chairman S., Charvex, CheesenibIets69, Cheesyk1991, Chick Bowen, Chimin 07, Chowbok, Chris 73, Chris G, Chris the speller, Chriskhong, Christadelphos, Christian List, Christina Silverman, Christopherlin, Cissy15, CiudadanoGlobal, Clavecin, ClementSeveillac, Closedmouth, CloudNine, Cntras, Cococathy105, Coffee, Colonies Chris, Commander, CommonsDelinker, Conversion script, Coolkate1, Coolmanedward, Copysan, Corporal Tunnel, Corpx, Corvus coronoides, Cosmic Latte, Cosprings, Courcelles, Craig Stuntz, CraigSilver, Cricket02, Crochet, CrookToe, Crstewartjr, Crunchy Numbers, Crushcrushcrushh, Ctbolt, Cucucuconn, D, D Hill, DARTH SIDIOUS 2, DCGeist, DJ Clayworth, DJTCUTZ, DSFanatic, DVD R W, DVdm, Dacookiemunster, Damirgraffiti, Dan hardboy, Dancey2, DancingPenguin, Dane Sorensen, Daniel Fenn87, Daniel Newman, Daniel913, DanielD1980, Danielbb2013, Danmayna, Danno uk, Danthur, DarkLover93, Darkwhistle, Daviticus82, Dawnseeker2000, Dbolton, Deadworm222, Deagle AP, Dekisugi, Delicious carbuncle, Delusion23, DerHexer, Desiromeo107, Devper94, Dforest, Dgillett, Dgw, Diannaa, Dice326, Dinnerbone, Dirgni1986, Discospinster, Dismas, Dlohcierekim, Dmaiolo, Dmcc, DocWatson42, DocendoDiscimus, DogcatcherDrew, Dogman15, Doktor Faustus, Doublebassistmagazine, Download, Dpotter, Dr. Friendly, Draconx, DragonflySixtyseven, Dragonzaid6, Draicone, Dreadstar, Drmies, Drondent, DuGly1, Ducknish, Duinemerwen, Dumbhead666, Dureo, Dust Filter, Dustyrannar123, Dygituljunky, Dysprosia, E0steven, ESkog, EWS23, EWikist, East718, Edderso, Edison, Ednerdy, Edward321, Edwardywang, Edwin M. Good, Ejeinowski, El C, EldKatt, Emil76, Emilyabbimayson1, Enviroboy, Epbr123, Equendil, Erathgirl56, EricSerge, EricWesBrown, Erielhonan, Erpert, Esslk, Etincelles, EugeneZelenko, Euryalus, Evan.willis, Evanreyes, Everyking, Evil saltine, Excirial, Exir Kamalabadi, Eziekiel Summers, FAIRYTAILAWESOMENESS, Fairoozrules12345, Falcon8765, Fanoftheworld, FarleyWiki, Femto, Fenevad, Fennec, Fetchcomms, Fieldday-sunday, Firepitluke, Firien, FisherQueen, Flameeyes, Flewis, Fluence, Foodisgood123, Forever Dusk, Formation, Freakmighty, Fredrik, Frehley, FreplySpang, Friedrich von Königsberg, Frosty0814snowman, Func, Funllama680, FunnyYetTasty, Furries, GHe, Gaidheal1, Gail, Gakrivas, Gala4ed, Gary King, Gawaxay, Gazimoff, Gbag09, Georgia guy, Gex999, Giftlite, Gilderien, Gilgamesh, Giovannibomoll, Giroper, Glane23, Globalsolidarity, Gniw, Gnomon, Gnowor, Goatselover22, Goawayforeveru, Gogo Dodo, Goodnightmush, GorillaWarfare, Gott34, Graham87, Greenshinobi, Gregmasterson, GreyCat, Grunty Thraveswain, Gryffindor, Guest9999, Gunmetal Angel, Gurch, Gwendal, Gwernol, Gymnopedist, Hadal, HaeB, Haearnbran, Hagie, Hall Monitor, HangingCurve, Hannahkkk, HappyInGeneral, HappyLogolover2011, Harland1, Haroldcamping, Harrylegendkitty, Hazmat2, Hdt83, Head, Heather998, Heggy, Hello19971997, Hellounable123, HenryLi, Hephaestos, Herbythyme, Hernelodge, HexaChord, HiDrNick, Hilmarleujes, Hobartimus, Hoof Hearted, Hoogoo, Houtlijm, Husond, Hwjh sk8er, Hyacinth, Hydnjo, II MusLiM HyBRiD II, IMPERIAL, IRP, ISBN, Iamcool1234567890, Icairns, IceUnshattered, Ida Shaw, Ilikethesnowcozitscool, Illuminatedwax, Iloveoliver, Ilovepiano, Imprimez, InFairness, Indon, Infrogmation, Intelati, InternetMeme, Iridescent, IronGargoyle, Ishmayl, It200, ItsZippy, Ixfd64, J d hurley, J. Van Meter, J.delanoy, J991, JAL78, JForget, JHCC, JJ Harrison, JNW, Jack1755, Jack1993jack, Jackehammond, Jackjack3000, JackofOz, Jackol, Jacob123321, Jadthedestroyer, Jagged 85, Jailane, JakeOneball, Jakejericho94, Jakew, Jamesontai, Jan1nad, Jaquanzxc, Jasper Deng, Jay Litman, Jazzpno, Jbusenitz, Jefchip, Jeffdougieg, Jello is yummy, Jengirl1988, Jenniferflynn89, JeremyA, Jerome Kohl, Jesta, Jfpierce, Jh51681, Jim1138, Jiy, Jjron, Jk2q3jrklse, Jkt46, Jlking3, Jmundo, JodyB, JoeSmack, Joelthesecond, Joemacgames1, Johann Wolfgang, John K, Johncohrs, Johnkarp, Johnnybb, Jojit fb, Jonathantrengrove, Jongleur100, Jonmaveni, Jonpaulusa, Jose77, Josh3580, Joshkorpolinski, Jossi, Jovianeye, Jpceayene, Jprg1966, Jpsousa4, Jpuff19, Jritche2, Jsa23899, Jswhit02, JuJube, Jubileeclipman, Julesd, Julia Rossi, Jump123, JuneGloom07, Jusdafax, Just plain Bill, Justor15, Juzeris, Jwolfe, Jwpurple, KANNIBAL, Ka Faraq Gatri, Kaihsu, Kaitlyn77, Karada, KarateKid96, Karenjc, Karol Langner, Kat, Kate, Katieh5584, Kayla.Cooper, Kbh3rd, Kbthompson, Kcordina, Keakealani, Keegan, Keilana, Kelovy, Keron Cyst, Ketsuekigata, Kevinmon, Killing Vector, Kinaro, King of Hearts, Kingpin13, KirinX, KleptHoeManiac, Knezzy9, Knucmo2, Knutux, Krawi, Krenakarore, Kristen Eriksen, Kubigula, Kurtguz, Kuru, Kvng, Kxug1234, Kystephkwan3, L Kensington, LAX, LOL, La Pianista, LadyKeelah, Lakeoftea, Lamellama, Lankiveil, Lasjenika daff, Latitudinarian, Latulla, Lawl95, Lcarus27, Ld, LeaveSleaves, LedgendGamer, LeilaniLad, Leonard G., Lerdthenerd, Lethe, Leuko, Liarliar2009, LibLord, Librarian2, LibraryLion, Liftarn, Lightfagbolt, Lightningstar101, Lights, Limefrost Spiral, Liquidblue8388, Little Mountain 5, LittleOldMe, Littlemissnikki1995, Livajo, Lklena, Logan, Loginnigol, Lom Konkreta, LoneliestMaster, Longhair, Look Busy, Lookingforgroup, LorenzoB, Losan33, Lothar Klaic, Lozeldafan, Lph, Lradrama, LshIII, Luckas Blade, Lucy6567, Luk, Lukas²³, LukeSurl, Lumos3, Luna Santin, Lupin, Lzur, M ajith, M here 4 ever, M0rphzone, MACKA004083, MC MasterChef, MER-C, MHLU, MJ94, MKleid, MONGO, MSGJ, Madhero88, Mafal, Magister Mathematicae, Mailgkhandelwal, Majorminormusic, Makerman123, Malafaya, Malcolmxl5, Mandel, Mandolinface, Manga-kid, Mangojuice, Mani1, Marek69, Mark Foskey, Markus liverpool, Martin451, Massysett, Materialscientist, Matt Deres, MatthewCMehrtens, Matthewrbowker, MauriceJFox3, Mawkish1983, Maxmoefoe1, Maxxo, MearsMan, Mediran, Meegs, Melodia, Melodramatically, Merphant, Mes0h0od, Mesolimbo, Mgrand, Mhking, Michael93555, Michaelsbll, Michelle, Midhart90, Mike R, Mike Rosoft, Mikerrr, Mild Bill Hiccup, Mindmatrix, Minimac, MinorProphet, Miranda, Mireut, Miss Madeline, Missmarple, Mivio, Mjroots, Mmedina15, Mmxx, ModWilson, Modernist, Moggie2002, Momo122yeah, Monkeydung95, Monteros, Montrealais, Moodie507, Moogoogaipan, Moralis, Mordomo, Moreschi, Morgan12345678910, Morn, Moskvax, Mr Stephen, Mrmoocole, Mschlindwein, Msteveson, Mstroeck, Mtoto mdogo, Music Sorter, MusicalAds, Musser, MutantRockGod, Mwanner, Mww113, My family123, My2cents, Myanw, NCurse, NDCompuGeek, NYKevin, Naddy, Nakon, Namco235, Natalie Erin, Natural Cut, NawlinWiki, Nazkaylad, Neier, NellieBly, NewEnglandYankee, NewWaveKid, Newportm, Nguyenngaviet, Nick, Nick Number, Nikita Beriozkin, Nikkimaria, Nimavojdani, Nk, Nlu, Noah Salzman, Noblestood, Noetica, Nomenclator, Nono64, NorCal Bosie, NostinAdrek, Number Googol, Numbermaniac, Obelix83, Oda Mari, Odie5533, Ohnoitsjamie, Oldenergy, Oliver202, Olivier, Omicronpersei8, OnBeyondZebrax, OnePt618, OpenInfoForAll, Opus33, Orange Suede Sofa, Orangesandham, Organic Cabbage, Orphan Wiki, Oscar45596524, Ospalh, OttawaAC, Outoftunepiano, Outriggr, OverlordQ, Oxymoron83, P0mbal, P36ad, PTJoshua, Paige.Banaszak, Pamri, Pandrews76, Panpog1, PashaTarsius, Paul Erik, Paul-L, Pavel Vozenilek, Pax:Vobiscum, Pdcook, Pedromd, Peoplefromarizona, Pepper, PericlesofAthens, Permose, Perry Hotter, Perrypierce10, Persian Poet Gal, Peruvianllama, Peter Karlsen, Petertomas, Peteweez, Pfalstad, Pfctdayelise, Pgk, Phantomsteve, Pharaoh of the Wizards, Philip Howard, Philip Trueman, PhilipMW, Phso2, Pianiccisimo, Piano Play It, Piano non troppo, PianoNanny, Pianodude87, Pianogen, Pianolover1, Pianoman199, Pianoman315, Pianoplonkers, Pianosxxi, Picklepopper, Pie Man 360, Pinethicket, Pipian, Pippitherabbit, Pkarsch, Pladask, Pleasantville, Poindexter Propellerhead, PookeyMaster, PranksterTurtle, Prasadbrg, Press Start, Prestonmag, Private Butcher, Prof.rick, Prof.smith10101, Professor N, Professor marginalia, Provedadwrong, Psy guy, Publiceyemedia, Purveyor of Great Knowledge, PuzzletChung, Pyrospirit, Qaqaq, Qartis, Qtoktok, Quale, Quantumobserver, Quelasol, Quiescen, Quintote, QuiteUnusual, Qwe, Qwerwqerwqer1, Qxz, RA0808, RJaguar3, Rainwarrior, Rami R, Randomized01234, Rashad9607, Ratiocinate, Raudys, RayquazaDialgaWeird2210, Razorflame, Rconroy, Rdouillette, Reaper Eternal, Rebekah2, Redrapidsreadingrats, Regancy42, Reggina, Regibox, RenamedUser01302013, Res2216firestar, Retired username, Rich Farmbrough, RichardMills65, Rigadoun, Rigaudon, Rikkiapples, RiseAbove, Rj1200, Rjbalon, Rmhermen, Ro2nie, Robchurch, Robert4565, RobertG, RobertMel, Robertfisner, Robshill, Rockhopper10r, Rod Corkin, Rodrigo Schejtman, Ronhjones, Roo60, RoosterBubble82, RoseSoul, Rothorpe, Rparucci, Rror, Rsrikanth05, Rtucker913, Ruandpiano, Ruthiedee, Rverne8, Ryan032, Ryenocerous, S1ckcacoon, SDS, SEWilco, SFK2, SJP, SQGibbon, STV0726, Salmar, Saltlakejohn, Sam Korn, Sandstein, Sango123, SarekOfVulcan, Satellizer, Scarian, Scariap, Sceptre, Scetoaux, Schafferdude18, SchfiftyThree, Schissel, SchreiberBike, Schwnj, Sciurinæ, Scotland-inch, Scotwriter, Scythre, Seaphoto, Sebih1, Sejunekwon, Sentausa, SeoMac, Sesshomaru, Seviolia, Sfhintern, Shanes, Shawnc, SheepNotGoats, Shem1805, Shirik, Shirulashem, Shizhao, Shoehornian, Shorne, Shortenfs, Shpakovich, Silent reverie86, Simatbirch, SimonP, Sintaku, Sir Dejan Djordjevic, SirEuropeLads, SirVulture, Sk8er517, Skarebo, Skittlesroxy111, Skizzik, SkyMachine, Skylarbblue, Slightsmile, Sluzzelin, Smellltheeglove, Smjg, Soccerpianoluvver, Soerfm, Solarra, Some jerk on the Internet, Soundscapemusic88, SpaceFlight89, Sparafucil, Sparcloud, Spartacan, Spartan, Special-T, Specs112, Spellcast, Spettro9, Spiesr, Spitfire, Spondoolicks, Springeragh, Sstrader, StAnselm, Staffwaterboy, StaticGull, Stephanie95519, Stephaniechung, Stephenb, Steve Bob, Steve2011, Stevekeiretsu, Stradfan, Strangerer, Stylishpony123, SudoGhost, SudoGhost (Away), SupaStarGirl, Superchick106, Supercko, Svizra, Sweetpoet, Swhidden, Swiftblade21, Switchercat, Sydneyrocks247, Synthdreamer, Syphon8, T-borg, THD3, THEN WHO WAS PHONE?, TMC1221, TYelliot, Tablizer, Taeshadow, Tagishsimon, Tammy Tang, Tarquin, TastyPoutine, Taurrandir, Tavakoli543, Tayste, Tbhotch, Tcncv, Tdadamemd, Teaghan11, Technopat, Ted178, Ted87, Teenwriter, Tenebrae, TeppoM, Teptoria, TerranG3, Tex, TexasAndroid, Thanrattonkansit, The Bread, The High Fin Sperm Whale, The Rambling Man, The Thing That Should Not Be, The juggresurection, The undertow, ThePirate, TheScotch, Thecazyandstupidone, Thecheesykid, Themeparkphoto, Themfromspace, Theo Buckley, Thereoncewasafarmerwholivedonarock, Theweekly, Thingg, Think outside the box, Thousandandone, Tide rolls, Tim Starling, Timneu22, Timrem, Tmbg lover, Tmopkisn, Tntrusty17, Tnxman307, Todeswalzer, Tolly4bolly, Tom, Tomaxer, Tombomp, Tomcat7, Tommyjb, Tommyyu22, Toon Link 891, Toughpigs, Tpbradbury, Tree Biting Conspiracy, Trevor MacInnis, TripLikeIDo, Triplestop, Troyeebarua, Trusilver, Tubby, Tucker001, Tumblingboulder, Tungchit, Turlo Lomon, Turnstep, Tuvwxyz, Tuyvan, TyA, Tyguy123, Uirauna, UncleAndyBob, UncleBubba, Unfree, UnfriendlyFire, Unint, UninvitedCompany, Unused0022, UomoDiHonore,

Upclicks, Urbanguy1, V.narsikar, V44sandy, Vanished user 39948282, VegaDark, Veinor, Vercalos, Versus22, Vicenarian, Viewwiki, Vilcxjo, Violncello, Vrenator, Vslashg, Vsmith, WJBscribe, Wakted152, Walter Humala, Wangi, Wanwan, Wavelength, Wayne Slam, WayneConrad, Wayward, Webclient101, Weed88980, Wengrobin, Werldwayd, WhaleyTim, WhiteOak2006, Whkoh, Who.was.phone, WickTr, Wiki alf, WikiLaurent, Wikibob250, Wikidudeman, Wikieditor06, Wikiname1109, Wikipelli, Wikiwakiwaki, Wilhelm666666, Willi Gers07, WilliamSommerwerck, Windchaser, Wisdom89, Wisteriapress, Witchwooder, Wizard191, Wjbean, Wolfdog, Wrestplank, Wtmitchell, Ww, X!, XEZx4Shot, XXBobertXx, Xaltoi, Xbrebre15, Xin-Xin W., Xtfcr7, XxMrs.KerestusXx, Yamahamusician, Yamamoto Ichiro, Yamla, Yaylorclevenger, YellowMonkey, Yeyonghe, Yoos2me, Yopogono, Zachary.meier, Zachcat, Zenohockey, Zenophilius, Zeroman2468, Zimbricchio, Znethru, Zoicon5, ZooFari, Zsinj, Zvika, Zvn, Zzxy69, Ásgeir IV., Ô, Саша Стефановић, Яippawallet, தமிழ்க்குட்டி, 2747 anonymous edits

Francis_Poulenc *Source*: http://en.wikipedia.org/w/index.php?title=Francis_Poulenc *Contributors*: 19DrPepper91, 97198, Abrashkin, Addaick, Ahoerstemeier, Aitias, Akriloth2160, Alfietucker, Amog, Andres, Andrew Robinson London, Andycjp, Antandrus, Aristote33, Arthena, Arturo57, Bbb23, Bearcat, Bgraabek, BillFlis, Bills16309, Bluetooth954, Bobo192, Boleyn, Bornhj, Camembert, Causa sui, CharlotteinWeimar, Charvex, Classickol, Cleduc, Cmulrooney, Cnoel77, Crzrussian, Cstaffa, D6, DChapii, DO'Neil, DTOx, DW, Danmuz, Date delinker, Dave314159, David Hoeffer, David Martland, David Underdown, DavidRF, Deb, DevAlt, Diotti, Dodoïste, Dori, DrKiernan, Dybryd, Ed g2s, Eliz81, ErinKM, FNawrocki, Flamurai, Fluffernutter, Fourthpoliceman, Francis Schonken, Frglz, Fwb44, Gimmetrow, Goldenlane, Goodmanjaz, Gretab, Grstain, Guanaco, Hadal, Hangsuwan, Harrypeterfrog, HenryLarsen, Hephaestos, Housewatcher, Hyacinth, Iwavns, J.delanoy, JackofOz, Jaysuc, Jerome Kohl, Jerzy, JillandJack, JmCor, Juliancolton, Jwillbur, JzG, Karamveer 93, Kelovy, Kenny Gill, Kierkkadon, Kleinzach, Koavf, Korinth111, Kray, Krun, Kumioko (renamed), Kwamikagami, Laonikoss, Lazulilasher, Leonard Vertighel, Leopoldhw, Linkoman, Luke1029, Luna Santin, Lzur, Magioladitis, Makemi, Mardetanha, MarkSutton, Martin S Taylor, MattDawg579, Mayumashu, Midway, Miracle Pen, Missmarple, Mordant21, Musikfabrik, Muu-karhu, Mycroft1854, Nixeagle, Nono64, Nrswanson, Nufftin, Oakshade, Oconnell usa, Of7271, Omnipaedista, Orbicle, Otto Jongerius, Pabix, Parkwells, Per Honor et Gloria, PerryTachett, Petrus Viljoen, PiCo, Piano non troppo, Poccil, Postdlf, Potatoscone, Pymouss, Rc65, Richard cocks, Rigaudon, Robert.Allen, RobertG, Robertgreer, Robertwalton, RodC, Roger Davies, Rothorpe, Ruth-2013, Sallyrob, Santiparam, SatyrTN, Sdr, Seann, Ser Amantio di Nicolao, Sj, Sketchee, Socky22, Som1950, Squeezeweasel, StaticGull, Stepshep, Stirling Newberry, TAnthony, Tassedethe, Tbhotch, Ted Wilkes, The Thing That Should Not Be, Timneu22, Tjmayerinsf, Tomaxer, TrollVandal, UncleBubba, UrbanGrill, Useight, UtherSRG, V79benno, VMS Mosaic, Varnent, Vega2, Veggieburgerfish, Violncello, Vrenator, Wikipelli, Woohookitty, Yankeecook, Zoe, Žiga, Александр, Андрей Романенко, Ὁ οἶστρος, 269 anonymous edits

C_major *Source*: http://en.wikipedia.org/w/index.php?title=C_major *Contributors*: A. Parrot, Aaronrp, Amber388, Andeggs, Andyzweb, Anton Mravcek, Ashthud, Bakpak2hvy, Bdesham, Bigbluefish, Bisqwit, Bluesky2004sky, Bongwarrior, Brittany Ka, BurnDownBabylon, Cbsteffen, Chekaz, Classicalfan2, Cloverboy19, Colonies Chris, CommonsDelinker, Cosh, Cst17, Daniel Newman, Denelson83, Dmetric, EditorE, Eric444, Evil saltine, Francs2000, Frigoris, Gamsbart, Georgia guy, GoingBatty, Grantorino70, Guitar chords, Guitardemon666, Gunsnroads, Horn of Plenty, Hyacinth, Ixfd64, JackofOz, James599, Jindřichův Smith, Johann Grimm, Johnny Au, Joseph Solis in Australia, Kingboyk, Kratos of Minnesota, Kyem2010, Linker0987654321, M.J.E., Mahlerlover1, Mapsax, Marcus2, Marie de France, Mathilde de la Mole, MegX, Mego'brien, Melsaran, Mgard7331, MiPe, Michael Angelkovich, Mr.BipolarCorner, Nick, NikNaks, O.Koslowski, ONEder Boy, PeterBFZ, Phil Sandifer, Piano non troppo, Plastikspork, R'n'B, RedJoker25, Rigadoun, RiseAbove, Robsteadman, Rockyou12, Ross280, Rossyxan, SQGibbon, Satchfan, SchfiftyThree, Schweiwikist, Ser Amantio di Nicolao, Simonsa, Sobreira, Spartan2005, SpinyMcSpleen, Springeragh, Sw607813, Tarret, Teiladnam, Themfromspace, Theresa123123123, Timothy Gu, Toddles29, Torc2, Vayu, Veggieburgerfish, Violncello, Volunteer Sibelius Salesman, Whiteghost.ink, Wikid77, Yms, Ã◆£, 205 anonymous edits

B-flat_minor *Source*: http://en.wikipedia.org/w/index.php?title=B-flat_minor *Contributors*: 54together, Ace of Spades, Amaury, Amber388, Andeggs, Andros 1337, Anton Mravcek, Aphasia83, C777, Cbsteffen, Classicalfan2, Cloverboy19, Cst17, DTOx, Damian Yerrick, Dekyl9, Double sharp, Dugwiki, EscapingLife, Extraordinary Machine, Flisdaswis, Guitar chords, Gunsnroads, Hirohisat, Hyacinth, Iamaneyeblinkernoseblower, Ianweller, JYi, JackofOz, James599, K1Bond007, Kbog, Kingturtle, La Pianista, Lanthanum-138, Lionhead99, M.J.E., Mahahahaneapneap, Marcus2, Marie de France, Martarius, Mgard7331, Michael Angelkovich, Michael Bednarek, Ndrly, Phil Sandifer, Plastikspork, Pr1268, Prabash.A, Qfiggly, Qwyrxian, Razorhead, Rigaudon, Rl, Robert Happelberg, Roshan220195, Rossyxan, Satchfan, Sobreira, Soccer-holic, Springeragh, Tegrenath, Themfromspace, Theopolisme, Theresa123123123, Tim!, Torc2, Ucucha, Uncle G, Violncello, Volunteer Sibelius Salesman, Wikid77, Yauvova Graffat, 130 anonymous edits

E_minor *Source*: http://en.wikipedia.org/w/index.php?title=E_minor *Contributors*: A. Parrot, Airplaneman, Amcbride, Andeggs, Andrewlargemanjones, Anton Mravcek, Aphasia83, Aspects, Aux2master, Berimbau1, Bigbrainkt, Bongwarrior, Cbsteffen, Chasedraney, Coffeezombie, ColdACARMINE, Crazydrummer2007, Cst17, Deltabeignet, Diloretojazz, Djproject, Dmetric, Doctorknock, Dwarf Kirlston, EgbertMcDunk, Flisdaswis, Fredil Yupigo, Frigoris, Georgia guy, Gryspnik, Guitarguy01, Gunsnroads, Hyacinth, Ideal gas equation, JackofOz, Jafeluv, James599, KelleyCook, Keycard, Krashlandon, Lambiam, Lanthanum-138, Marcus2, Martarius, Marvin Chinchilla, Mathpianist93, MegX, Methegreat, Mgard7331, Michael Angelkovich, MonoAV, Mr bouregaurd, Musicmaster345, Ndrly, Nkieper, Phil Sandifer, Plastikspork, R.christie, RafaAzevedo, Razorhead, Redstrad, Richhoncho, Rossyxan, Rotteth, RoyBoy, Satchfan, SchfiftyThree, Serenitynoire, ShelfSkewed, Sobreira, SpinyMcSpleen, Springeragh, Spydrfish, Tassedethe, Tide rolls, Toddles29, Torc2, Violncello, Volunteer Sibelius Salesman, Wasted Time R, Whiteghost.ink, Whitepaw, Wikid77, Wild Matt12, Wolfer68, Woohookitty, Zeisseng, Ã◆£, 212 anonymous edits

Manuel_de_Falla *Source*: http://en.wikipedia.org/w/index.php?title=Manuel_de_Falla *Contributors*: 0000000000000, 4meter4, Abudwig, Addshore, As;akjf, Auntof6, Ayrton Prost, BanyanTree, Bernstein2291, Bjankuloski06en, Btw, Camembert, Chris the speller, Claudioarrau, CutOffTies, D6, DanielVonEhren, Deb, Deborahjay, Dionisis1979, DrG, Dslhgfdw, Dysepsion, EastTN, Ebakunin, Egemont, Elfelix, Etincelles, Europus, Everyking, Feline Hymnic, Flamurai, Fresnomo, Gadfium, Gaudio, Geremia, Gian-2, Graham87, GuillaumeTell, Gurch, Gyan, Homagetocatalonia, HombreDHojalata, Hu12, Hyacinth, Ilizaran, Impy4ever, Introvert, Island Monkey, J.delanoy, JackofOz, Janstoneman, Japanese Searobin, JustAGal, Karl Stas, Karljoos, Kbwiki001, Kleinzach, Laonikoss, Liftarn, Limideen, M7, Makemi, Marcus2, Melodia, MichaelHaeckel, Missmarple, Monegasque, Obradovic Goran, Paul A, Posible2006, Pratyya Ghosh, QwertyUSA, Raul654, Rje, Robertgreer, Singingdaisies, Smec, Stahlbrand, TBHecht, TJRC, Tatuaje, Technopat, The Determinator, The Thing That Should Not Be, The Utahraptor, Thingg, Timneu22, Timothykirk, Tjmayerinsf, Tomaxer, UninvitedCompany, UtherSRG, Voceditenore, VoroSP, Vstrad7, Warrickball, Wii 2, Wikiwikiwaahwaah, 148 anonymous edits

Time_signature *Source*: http://en.wikipedia.org/w/index.php?title=Time_signature *Contributors*: 2602:306:2507:3529:71AF:4897:C451:ED2E, 2620:C6:8000:800:1293:E9FF:FE0F:4A14, 2A01:E35:8A55:A580:C0E7:3424:5D55:5CB7, A Softer Answer, A2Kafir, A4, AOC25, Adam78, Addshore, Alansohn, Alex299006, Alexf, Allansteel, Alpha Quadrant (alt), AnArbitraryConstant, Andrewpmk, Angr, Antandrus, Antediluvian67, Antti Salonen, Aquae, Arcadian, Arjun01, Arthena, Atif.t2, Avrv, AzaToth, Baccyak4H, Barneca, Bdesham, Beeblebrox, Bemoeial, Ben Culture, Ben Tibbetts, BenFrantzDale, Beofluff, BeteNoir, Bgamari, Bkell, Bobo192, Bonalaw, Branddobbe, Bshway, Bt8257, Burnside65, CSWarren, Caesura, Caltas, Camembert, Cfoughty, Chewie, ChrisGualtieri, Chrishmt0423, Christopher Parham, Chuckiesdad, Chuckthekicker, Conversion script, Cyberkid ua, Dachshund, Damaged again, Damian Yerrick, Dark Silver Crow, Dave Taylor, Dcamp314, Deckiller, Defdoc, Dekimasu, DeltaQuad, Deltabeignet, DimebagJr., Djcountessb, Dkaplowitz, Doc Strange, DocWatson42, Donarreiskoffer, Dreamyshade, Drummer070, Drums4sure, DuncanHill, Dureo, Dysprosia, Dze27, ESSch, Edgaropolous, Ellsworth, Eminemfan342, Emmett5, Emperorbma, Epbr123, Erasmussen, ErkinBatu, Everyking, Evil Monkey, Evil saltine, Faustohmj, Feline Hymnic, Ferdinand Pienaar, Florian Blaschke, FlyHigh, Forbes72, FordPrefect42, ForgeGod, Freedomlinux, Furrykef, Gekritzl, Genes6, Georgia guy, Gershom, Giftlite, Gilderien, Gmcole, Goosegoosegoose, Graham87, Grendelkhan, Guocuozuoduo, H2g2bob, Haberg, Hairy Dude, Heavenlier, Hig Hertenfleurst, HighHopes, Hmains, Husseinmoien, Hyacinth, ILike2BeAnonymous, Iamunknown, Ijon, Impy4ever, Iridescent, It Is Me Here, J.delanoy, J04n, J424, JForget, Jakarr, James599, Jashiin, Jerome Charles Potts, Jerome Kohl, Jh51681, Jkeuler, Jocke666, Joel D. Reid, Jogloran, Jordanotto, Jorgicio, Joseph Solis in Australia, Josh Parris, Joshk, Joshua Boniface, JoshuaGarton, Jtle515, Jubileeclipman, Jusses2, KathrynLybarger, Kbh3rd, Kemokid70, Kevds318, Kf4bdy, Khullah, Kitch, Kooo, Korona, Koyaanis Qatsi, Lamontmcpheron, Lanthanum-138, Lawstubes, LeaveSleaves, Leszek Jańczuk, LinDrug, Lmatt, Looper5920, MMBKG, MONGO, Macgruder, Madhero88, Marblespire, Martarius, MatrixFrog, Mav, MegX, Mel Etitis, Melonkelon, Mernen, Merphant, Mervyn, Mets501, Mikeblas, Missmarple, MlI1013, Mordomo, Muhandes, Munci, Munin, MuzikJunky, Mysid, N64349, Naddy, Nahum, Neumeier, Neutrality, NewEnglandYankee, Nibble, Nihil, Nobody can see me, Nsaa, ONEder Boy, Oda Mari, OldakQuill, Omegatron, Oodles More, Opus33, Oscillon, Oxinabox, Oxymoron83, Panzak7, Papley, Pasajero, Patstuart, Pelarmian, Per84, Pewwer42, Philip Howard, Phlake, PochWiki, Pt, Ptrue, Qintinging, Quadalpha, Quentin X, Quux, R Lowry, R. fiend, RScheiber, RadioFan, RafaAzevedo, Ratzer, RedWolf, Redd Foxx 1991, Reisio, Retired username, RexNL, Rigaudon, Rizniz, Rjd0060, Rjwilmsi, Rmhermen, Robertinventor, Ronhjones, ST47, Sadalmelik, Schissel, Schyler, Shadowjams, ShelfSkewed, Simon the Likable, SimonEast, Skizzik, Skusek, Skytopia, Smartie nerdy, Smyth, Snowy150, Sparafucil, Squandermania, Stannered, Stapler 9 42, Starwiz, Steven Zhang, Stratford15, Stu57, Stwalkerster, Suddenly There Is a Valley, Sunray, Super Rad!, Svick, Svm2, Synchronism, T3arstain, Tahren B, Tarquin, TashTish, TenPoundHammer, The Evil IP address, The Thing That Should Not Be, The wub, TheDoober, Timsabin, TonySt, Tpbradbury, TrebleClef, TrevorX, Trevordeclercq, Trlkly, Ufwuct, Universalcosmos, Userresuuser, Utcursch, Vanished user 39948282, VdSV9, Violncello, Volunteer Sibelius Salesman, Vrenator, Wahoofive, Waldir, Wereon, Wik, WikiPBia, Wikid77, Wmattis, Wmpuig, Woohookitty, Wwwwolf, Xinoph, Yanksox, Zephalis, Zundark, 699 anonymous edits

El_amor_brujo *Source*: http://en.wikipedia.org/w/index.php?title=El_amor_brujo *Contributors*: 4meter4, AlbertSM, Andycjp, Ary29, BomBom, Colonies Chris, Dark Silver Crow, DavidRF, Egemont, Ekem, Etincelles, Focus, Fvasconcellos, Gidiav, HarryHenryGebel, JackofOz, Karljoos, Kjell Knudde, Kleinzach, Malcolma, Ogg, Parkwells, Pianoplonkers, Psantosj, QwertyUSA, Rich Farmbrough, Robertgreer, SatcMoug, SnowRaptor, Sojambi Pinola, TBHecht, TJRC, The Anome, Tomaxer, Ulric1313, Vstrad7, Woohookitty, 24 anonymous edits

Image Sources, Licenses and Contributors

Image:Grand Piano 1781 France - Louis Bas.jpg *Source*: http://en.wikipedia.org/w/index.php?title=File:Grand_Piano_1781_France_-_Louis_Bas.jpg *License*: unknown *Contributors*: User:Charvex

Image:FortepianoByMcNultyAfterWalter1805.jpg *Source*: http://en.wikipedia.org/w/index.php?title=File:FortepianoByMcNultyAfterWalter1805.jpg *License*: unknown *Contributors*: Original uploader was Opus33 at en.wikipedia

Image:Broadwood grand square action.svg *Source*: http://en.wikipedia.org/w/index.php?title=File:Broadwood_grand_square_action.svg *License*: unknown *Contributors*: User:Mireut

Image:Erard double pilot action.svg *Source*: http://en.wikipedia.org/w/index.php?title=File:Erard_double_pilot_action.svg *License*: unknown *Contributors*: Pierre Erard

Image:DuplexScale.JPG *Source*: http://en.wikipedia.org/w/index.php?title=File:DuplexScale.JPG *License*: unknown *Contributors*: Joel Lidstrom

Image:Upright piano inside.jpg *Source*: http://en.wikipedia.org/w/index.php?title=File:Upright_piano_inside.jpg *License*: unknown *Contributors*: User:Pko

File:Thelonious Monk, Minton's Playhouse, New York, N.Y., ca. Sept. 1947 (William P. Gottlieb 06221).jpg *Source*: http://en.wikipedia.org/w/index.php?title=File:Thelonious_Monk,_Minton's_Playhouse,_New_York,_N.Y.,_ca._Sept._1947_(William_P._Gottlieb_06221).jpg *License*: unknown *Contributors*: ALE!, Freimut Bahlo, Ras67

Image:Fortepian - schemat.svg *Source*: http://en.wikipedia.org/w/index.php?title=File:Fortepian_-_schemat.svg *License*: unknown *Contributors*: Made in cooperation with User:Bechstein.

Image:Piano in Entrance Hall.jpg *Source*: http://en.wikipedia.org/w/index.php?title=File:Piano_in_Entrance_Hall.jpg *License*: unknown *Contributors*: White House (Zhen-Huan Lu)

Image:Klavier nah offen.jpg *Source*: http://en.wikipedia.org/w/index.php?title=File:Klavier_nah_offen.jpg *License*: unknown *Contributors*: User:Manuel_Strehl

Image:Steinway piano - Duo-Art small.ogv *Source*: http://en.wikipedia.org/w/index.php?title=File:Steinway_piano_-_Duo-Art_small.ogv *License*: unknown *Contributors*: User:PleaseStand

Image:D274.jpg *Source*: http://en.wikipedia.org/w/index.php?title=File:D274.jpg *License*: unknown *Contributors*: KarlKunde

Image:Steinway grand piano - pedals.jpg *Source*: http://en.wikipedia.org/w/index.php?title=File:Steinway_grand_piano_-_pedals.jpg *License*: unknown *Contributors*: User:Fanoftheworld

Image:Pedal Mark 1.svg *Source*: http://en.wikipedia.org/w/index.php?title=File:Pedal_Mark_1.svg *License*: unknown *Contributors*: User:Hazmat2

Image:Pedal Mark 2.svg *Source*: http://en.wikipedia.org/w/index.php?title=File:Pedal_Mark_2.svg *License*: unknown *Contributors*: User:Hazmat2

Image:Pedal piano 1.JPG *Source*: http://en.wikipedia.org/w/index.php?title=File:Pedal_piano_1.JPG *License*: unknown *Contributors*: Aotake, Clavecin, Clusternote, Shoulder-synth

Image:Estonia klaver 3.jpg *Source*: http://en.wikipedia.org/w/index.php?title=File:Estonia_klaver_3.jpg *License*: unknown *Contributors*: Epp

Image:SoundboardBracesRibs.jpg *Source*: http://en.wikipedia.org/w/index.php?title=File:SoundboardBracesRibs.jpg *License*: unknown *Contributors*: Gede, Thuresson, Wst

Image:Fluegel-Rahmen.jpg *Source*: http://en.wikipedia.org/w/index.php?title=File:Fluegel-Rahmen.jpg *License*: unknown *Contributors*: Original uploader was Kassander der Minoer at de.wikipedia

Image:Steinway Grand Piano Iron Plates and Strings.jpg *Source*: http://en.wikipedia.org/w/index.php?title=File:Steinway_Grand_Piano_Iron_Plates_and_Strings.jpg *License*: unknown *Contributors*: User:Dmaiolo

Image:Moritz von Schwind Schubertiade.jpg *Source*: http://en.wikipedia.org/w/index.php?title=File:Moritz_von_Schwind_Schubertiade.jpg *License*: unknown *Contributors*: AndreasPraefcke, Infrogmation, LeastCommonAncestor, Lewenstein, Wst, 1 anonymous edits

Image:Piano tuner.jpg *Source*: http://en.wikipedia.org/w/index.php?title=File:Piano_tuner.jpg *License*: unknown *Contributors*: User: BarbD

File:Francis Poulenc & Wanda Landowska.jpg *Source*: http://en.wikipedia.org/w/index.php?title=File:Francis_Poulenc_&_Wanda_Landowska.jpg *License*: unknown *Contributors*: Unattributed

File:PoulencCommemorativePlaque.jpg *Source*: http://en.wikipedia.org/w/index.php?title=File:PoulencCommemorativePlaque.jpg *License*: unknown *Contributors*: User:PHGCOM

Image:Root of a major chord on C.svg *Source*: http://en.wikipedia.org/w/index.php?title=File:Root_of_a_major_chord_on_C.svg *License*: unknown *Contributors*: User:Timothy Gu

Image:C Major scale (up and down).svg *Source*: http://en.wikipedia.org/w/index.php?title=File:C_Major_scale_(up_and_down).svg *License*: unknown *Contributors*: User:Bdesham

File:B-flat natural minor scale ascending and descending.png *Source*: http://en.wikipedia.org/w/index.php?title=File:B-flat_natural_minor_scale_ascending_and_descending.png *License*: unknown *Contributors*: User:Hyacinth

File:B-flat harmonic minor scale ascending and descending.png *Source*: http://en.wikipedia.org/w/index.php?title=File:B-flat_harmonic_minor_scale_ascending_and_descending.png *License*: unknown *Contributors*: User:Hyacinth

File:B-flat melodic minor scale ascending and descending.png *Source*: http://en.wikipedia.org/w/index.php?title=File:B-flat_melodic_minor_scale_ascending_and_descending.png *License*: unknown *Contributors*: User:Hyacinth

Image:E natural minor scale ascending and descending.png *Source*: http://en.wikipedia.org/w/index.php?title=File:E_natural_minor_scale_ascending_and_descending.png *License*: unknown *Contributors*: User:Hyacinth

Image:E harmonic minor scale ascending and descending.png *Source*: http://en.wikipedia.org/w/index.php?title=File:E_harmonic_minor_scale_ascending_and_descending.png *License*: unknown *Contributors*: User:Hyacinth

Image:E melodic minor scale ascending and descending.png *Source*: http://en.wikipedia.org/w/index.php?title=File:E_melodic_minor_scale_ascending_and_descending.png *License*: unknown *Contributors*: User:Hyacinth

File:Scale_deg_1.svg *Source*: http://en.wikipedia.org/w/index.php?title=File:Scale_deg_1.svg *License*: unknown *Contributors*: User:Mscuthbert

File:Scale_deg_2.svg *Source*: http://en.wikipedia.org/w/index.php?title=File:Scale_deg_2.svg *License*: unknown *Contributors*: User:Mscuthbert

File:Scale_deg_3.svg *Source*: http://en.wikipedia.org/w/index.php?title=File:Scale_deg_3.svg *License*: unknown *Contributors*: User:Mscuthbert

File:Scale_deg_4.svg *Source*: http://en.wikipedia.org/w/index.php?title=File:Scale_deg_4.svg *License*: unknown *Contributors*: User:Mscuthbert

File:Scale_deg_5.svg *Source*: http://en.wikipedia.org/w/index.php?title=File:Scale_deg_5.svg *License*: unknown *Contributors*: User:Mscuthbert

File:Scale_deg_6.svg *Source*: http://en.wikipedia.org/w/index.php?title=File:Scale_deg_6.svg *License*: unknown *Contributors*: User:Mscuthbert

File:Scale_deg_7.svg *Source*: http://en.wikipedia.org/w/index.php?title=File:Scale_deg_7.svg *License*: unknown *Contributors*: User:Mscuthbert

Image:ManuelDeFalla.JPG *Source*: http://en.wikipedia.org/w/index.php?title=File:ManuelDeFalla.JPG *License*: unknown *Contributors*: Bohème, Claudio Elias, 1 anonymous edits

File:Commontime inline.gif *Source*: http://en.wikipedia.org/w/index.php?title=File:Commontime_inline.gif *License*: unknown *Contributors*: User:Wikid77

File:Allabreve.svg *Source*: http://en.wikipedia.org/w/index.php?title=File:Allabreve.svg *License*: unknown *Contributors*: User:Joey-das-WBF

Image:common time signatures.gif *Source*: http://en.wikipedia.org/w/index.php?title=File:Common_time_signatures.gif *License*: unknown *Contributors*: A1, EDUCA33E, Fuebar, Juiced lemon, Wst, 1 anonymous edits

Image:3-4 equals 3-8 drum pattern.png *Source*: http://en.wikipedia.org/w/index.php?title=File:3-4_equals_3-8_drum_pattern.png *License*: unknown *Contributors*: User:Hyacinth

Image:12-8 equals 4-4 drum pattern.png *Source*: http://en.wikipedia.org/w/index.php?title=File:12-8_equals_4-4_drum_pattern.png *License*: unknown *Contributors*: User:Hyacinth

Image:Characteristic rock drum pattern.png *Source*: http://en.wikipedia.org/w/index.php?title=File:Characteristic_rock_drum_pattern.png *License*: unknown *Contributors*: User:SreeBot

Image:Half time rock pattern.png *Source*: http://en.wikipedia.org/w/index.php?title=File:Half_time_rock_pattern.png *License*: unknown *Contributors*: User:Hyacinth

Image:Simple duple drum pattern.png *Source*: http://en.wikipedia.org/w/index.php?title=File:Simple_duple_drum_pattern.png *License*: unknown *Contributors*: User:Hyacinth

Image:Simple triple drum pattern.png *Source*: http://en.wikipedia.org/w/index.php?title=File:Simple_triple_drum_pattern.png *License*: unknown *Contributors*: User:Hyacinth

Image:Compound duple drum pattern.png *Source*: http://en.wikipedia.org/w/index.php?title=File:Compound_duple_drum_pattern.png *License*: unknown *Contributors*: User:SreeBot

Image:Compound triple drum pattern.png *Source*: http://en.wikipedia.org/w/index.php?title=File:Compound_triple_drum_pattern.png *License*: unknown *Contributors*: User:Hyacinth

Image:Compound quadruple drum pattern.png *Source*: http://en.wikipedia.org/w/index.php?title=File:Compound_quadruple_drum_pattern.png *License*: unknown *Contributors*: User:Hyacinth

File:Video for 2o4 at 60 bpm.ogv *Source*: http://en.wikipedia.org/w/index.php?title=File:Video_for_2o4_at_60_bpm.ogv *License*: unknown *Contributors*: User:Robertinventor

File:Video for 3o4 at 60 bpm.ogv *Source*: http://en.wikipedia.org/w/index.php?title=File:Video_for_3o4_at_60_bpm.ogv *License*: unknown *Contributors*: User:Robertinventor

File:Video for 4o4 at 60 bpm.ogv *Source*: http://en.wikipedia.org/w/index.php?title=File:Video_for_4o4_at_60_bpm.ogv *License*: unknown *Contributors*: User:Robertinventor

File:Video of 6o8 at 90 bpm.ogv *Source*: http://en.wikipedia.org/w/index.php?title=File:Video_of_6o8_at_90_bpm.ogv *License*: unknown *Contributors*: User:Robertinventor

File:Video for 9o8 at 90 bpm.ogv *Source*: http://en.wikipedia.org/w/index.php?title=File:Video_for_9o8_at_90_bpm.ogv *License*: unknown *Contributors*: User:Robertinventor

File:Video for 12o8 at 90 bpm.ogv *Source*: http://en.wikipedia.org/w/index.php?title=File:Video_for_12o8_at_90_bpm.ogv *License*: unknown *Contributors*: User:Robertinventor

File:Video-of-5o4-at-60-bpm.ogv *Source*: http://en.wikipedia.org/w/index.php?title=File:Video-of-5o4-at-60-bpm.ogv *License*: unknown *Contributors*: User:Robertinventor

File:Video-for-7o4.theora.ogv *Source*: http://en.wikipedia.org/w/index.php?title=File:Video-for-7o4.theora.ogv *License*: unknown *Contributors*: User:Robertinventor

File:Video-of-11o4-at-60-bpm.ogv *Source*: http://en.wikipedia.org/w/index.php?title=File:Video-of-11o4-at-60-bpm.ogv *License*: unknown *Contributors*: User:Robertinventor

File:Blue-Rondo-à-La-Turk.theora.ogv *Source*: http://en.wikipedia.org/w/index.php?title=File:Blue-Rondo-à-La-Turk.theora.ogv *License*: unknown *Contributors*: User:Robertinventor

Image:Mussorgsky Pictures at an Exhibition, chords.PNG *Source*: http://en.wikipedia.org/w/index.php?title=File:Mussorgsky_Pictures_at_an_Exhibition,_chords.PNG *License*: unknown *Contributors*: User:Hyacinth

Image:Stravinsky, The Rite of Spring, Sacrificial Dance.PNG *Source*: http://en.wikipedia.org/w/index.php?title=File:Stravinsky,_The_Rite_of_Spring,_Sacrificial_Dance.PNG *License*: unknown *Contributors*: User:Hyacinth

Image:alternating time signatures2.gif *Source*: http://en.wikipedia.org/w/index.php?title=File:Alternating_time_signatures2.gif *License*: unknown *Contributors*: Original uploader was Wahoofive at en.wikipedia

File:Flamenco Compas.theora.ogv *Source*: http://en.wikipedia.org/w/index.php?title=File:Flamenco_Compas.theora.ogv *License*: unknown *Contributors*: User:Robertinventor

Image:additive time signature.svg *Source*: http://en.wikipedia.org/w/index.php?title=File:Additive_time_signature.svg *License*: unknown *Contributors*: User:Stannered

File:Video-for-time-signature-3+2+3.theora.ogv *Source*: http://en.wikipedia.org/w/index.php?title=File:Video-for-time-signature-3+2+3.theora.ogv *License*: unknown *Contributors*: User:Robertinventor

Image:orff time signatures.gif *Source*: http://en.wikipedia.org/w/index.php?title=File:Orff_time_signatures.gif *License*: unknown *Contributors*: Fuebar, Hyacinth, Wahoofive

File:Polymeter-4o4c4o3.theora.ogv *Source*: http://en.wikipedia.org/w/index.php?title=File:Polymeter-4o4c4o3.theora.ogv *License*: unknown *Contributors*: User:Robertinventor

File:Polymeter-2o6c3o4.theora.ogv *Source*: http://en.wikipedia.org/w/index.php?title=File:Polymeter-2o6c3o4.theora.ogv *License*: unknown *Contributors*: User:Robertinventor

File:Polymeter-2o5c2o3.theora.ogv *Source*: http://en.wikipedia.org/w/index.php?title=File:Polymeter-2o5c2o3.theora.ogv *License*: unknown *Contributors*: User:Robertinventor

Image:Mensural time signature 1.svg *Source*: http://en.wikipedia.org/w/index.php?title=File:Mensural_time_signature_1.svg *License*: unknown *Contributors*: Mysid

Image:Mensural time signature 2.svg *Source*: http://en.wikipedia.org/w/index.php?title=File:Mensural_time_signature_2.svg *License*: unknown *Contributors*: Mysid

Image:Mensural time signature 3.svg *Source*: http://en.wikipedia.org/w/index.php?title=File:Mensural_time_signature_3.svg *License*: unknown *Contributors*: Mysid

Image:Mensural time signature 4.svg *Source*: http://en.wikipedia.org/w/index.php?title=File:Mensural_time_signature_4.svg *License*: unknown *Contributors*: Mysid

Image:mensural proportion1.gif *Source*: http://en.wikipedia.org/w/index.php?title=File:Mensural_proportion1.gif *License*: unknown *Contributors*: (original uploader)

Image:mensural proportion2.gif *Source*: http://en.wikipedia.org/w/index.php?title=File:Mensural_proportion2.gif *License*: unknown *Contributors*: Wahoofive at en.wikipedia

Image:mensural proportion5.gif *Source*: http://en.wikipedia.org/w/index.php?title=File:Mensural_proportion5.gif *License*: unknown *Contributors*: Original uploader was Wahoofive at en.wikipedia

Image:mensural proportion4.gif *Source*: http://en.wikipedia.org/w/index.php?title=File:Mensural_proportion4.gif *License*: unknown *Contributors*: Original uploader was Wahoofive at en.wikipedia

MIX
Papier aus verantwortungsvollen Quellen
Paper from responsible sources
FSC® C105338

Printed by Books on Demand GmbH, Norderstedt / Germany